Photodetectors: Devices and Applications

Photodetectors: Devices and Applications

Edited by
Rowan Owens

Larsen & Keller
www.larsen-keller.com

Photodetectors: Devices and Applications
Edited by Rowan Owens
ISBN: 978-1-63549-219-4 (Hardback)

Published by Larsen and Keller Education,
5 Penn Plaza,
19th Floor,
New York, NY 10001, USA

Cataloging-in-Publication Data

Photodetectors : devices and applications / edited by Rowan Owens.
p. cm.
Includes bibliographical references and index.
ISBN 978-1-63549-219-4
1. Optical detectors. 2. Photodiodes. 3. Photovoltaic power generation. 4. Photoelectric multipliers. I. Owens, Rowan.
TK8360.O67 P46 2017
681.25--dc23

Printed and bound in the United States of America.

For more information regarding Larsen and Keller Education and its products, please visit the publisher's website www.larsen-keller.com

Table of Contents

Preface

This book elucidates the concepts and innovative models around prospective developments with respect to photodetectors. It talks in detail about the various concepts and theories related to this subject. Photodetectors refer to the sensors that sense light and electromagnetic energy. It has various types like polarization photodetector, photoemission photodetector, photovoltaic, and photochemical photodetectors, etc. The aim of this text is to equip students with the fundamental techniques and methods used in this area. Most of the topics introduced in the book cover new techniques and the applications of photodetectors. Through this book, we attempts to further enlighten the readers about the new concepts in this field. Coherent flow of topics, student-friendly language and extensive use of examples make this textbook an invaluable source of knowledge.

To facilitate a deeper understanding of the contents of this book a short introduction of every chapter is written below:

Chapter 1- Photodetectors transform light signals into voltage or current. Some of the examples of photodetectors are photodiodes and phototransistors. The frequency of these devices vary. The chapter on photodetectors offers an insightful focus, keeping in mind the subject matter.

Chapter 2- Responsivity, quantum efficiency, specific detectivity and dark current are some of the significant concepts of photodetectors. Responsivity measures electrical outputs per optical input whereas the merit used to measure the performance of a photodetector is known as specific detectivity. The following text unfolds its crucial aspects in a critical yet systematic manner.

Chapter 3- A photodiode helps in the conversion of light into electric current. Some of the applications of photodiodes are photoconductors, charge coupled devices and photomultiplier tubes. Common devices that have photodiodes are smoke detectors, remote controllers used to control televisions and air conditioner.

Chapter 4- The following text is an overview of the subject matter incorporating all the major aspects of photovoltaics. Photovoltaic cells convert light into electricity with the help of semiconducting materials such as silicon and selenium. This chapter also focuses on topics such as photovoltaic system, grid-connected photovoltaic power system, building-integrated photovoltaics and concentrator photovoltaics.

Chapter 5- This section helps the reader in developing an improved understanding of photomultipliers. Photomultipliers are extremely sensitive sensors of light and are members of the class of vacuum phototubes. Some of the subjects explained in this segment are photoreceptor cells, resonant-cavity-enhanced photo detectors, microchannel plate detectors and photoelectric sensors.

Chapter 6-There are many allied fields of photodetection technologies that use the basic concepts of photodetectors. These are technology whose functioning and process depends on the activation of light. Some common applications of photodetectors that are listed in this chapter are active pixel sensor, solar panels and solar cells.

I would like to share the credit of this book with my editorial team who worked tirelessly on this book. I owe the completion of this book to the never-ending support of my family, who supported me throughout the project.

Editor

1

Introduction to Photodetectors

Photodetectors transform light signals into voltage or current. Some of the examples of photodetectors are photodiodes and phototransistors. The frequency of these devices vary. The chapter on photodetectors offers an insightful focus, keeping in mind the subject matter.

Photosensors or photodetector are sensors of light or other electromagnetic energy. A photo detector converts light signals that hit the junction into voltage or current. The connection uses an illumination window with an anti-reflective coating to absorb the light photons. This results in creation of electron-hole pairs in the depletion region. Photodiodes and photo transistors are few examples of photo detectors. Solar cells are also similar to photo detectors as they absorb light and turn it into energy.

A photodetector salvaged from a CD-ROM. The photodetector contains three photodiodes, visible in the photo (in center).

Types

Photodetectors may be classified by their mechanism for detection:

- Photoemission: Photons cause electrons to transition from the conduction band of a material to free electrons in a vacuum or gas.
- Photoelectric: Photons cause electrons to transition from the valence band to

the conduction band of a semiconductor.

- Photovoltaic: Photons cause a voltage to develop across a depletion region of a photovoltaic cell.
- Thermal: Photons cause electrons to transition to mid-gap states then decay back to lower bands, inducing phonon generation and thus heat.
- Polarization: Photons induce changes in polarization states of suitable materials, which may lead to change in index of refraction or other polarization effects.
- Photochemical: Photons induce a chemical change in a material.
- Weak interaction effects: photons induce secondary effects such as in photon drag detectors or gas pressure changes in Golay cells.

Photodetectors may be used in different configurations. Single sensors may detect overall light levels. A 1-D array of photodetectors, as in a spectrophotometer or a Line scanner, may be used to measure the distribution of light along a line. A 2-D array of photodetectors may be used as an image sensor to form images from the pattern of light before it.

Properties

There are a number of performance metrics, also called figures of merit, by which photodetectors are characterized and compared

- Spectral response: The response of a photodetector as a function of photon frequency.
- Quantum efficiency: The number of carriers (electrons or holes) generated per photon.
- Responsivity: The output current divided by total light power falling upon the photodetector.
- Noise-equivalent power: The amount of light power needed to generate a signal comparable in size to the noise of the device.
- Detectivity: The square root of the detector area divided by the noise equivalent power.
- Gain: The output current of a photodetector divided by the current directly produced by the photons incident on the detectors, i.e., the built-in current gain.
- Dark current: The current flowing through a photodetector even in the absence of light.

- Response time: The time needed for a photodetector to go from 10% to 90% of final output.
- Noise spectrum: The intrinsic noise voltage or current as a function of frequency. This can be represented in the form of a noise spectral density.

Devices

Grouped by mechanism, photodetectors include the following devices:

Photoemission

- Gaseous ionization detectors are used in experimental particle physics to detect photons and particles with sufficient energy to ionize gas atoms or molecules. Electrons and ions generated by ionization cause a current flow which can be measured.
- Photomultiplier tubes containing a photocathode which emits electrons when illuminated, the electrons are then amplified by a chain of dynodes.
- Phototubes containing a photocathode which emits electrons when illuminated, such that the tube conducts a current proportional to the light intensity.
- Microchannel plate detectors are silicon-based photomultipliers.

Photoelectric

- Active-pixel sensors (APSs) are image sensors. Usually made in a complementary metal-oxide-semiconductor (CMOS) process, and also known as CMOS image sensors, APSs are commonly used in cell phone cameras, web cameras, and some DSLRs.
- Cadmium zinc telluride radiation detectors can operate in direct-conversion (or photoconductive) mode at room temperature, unlike some other materials (particularly germanium) which require liquid nitrogen cooling. Their relative advantages include high sensitivity for x-rays and gamma-rays, due to the high atomic numbers of Cd and Te, and better energy resolution than scintillator detectors.
- Charge-coupled devices (CCD), which are used to record images in astronomy, digital photography, and digital cinematography. Before the 1990s, photographic plates were most common in astronomy. The next generation of astronomical instruments, such as the Astro-E2, include cryogenic detectors.
- HgCdTe infrared detectors. Detection occurs when an infrared photon of suf-

ficient energy kicks an electron from the valence band to the conduction band. Such an electron is collected by a suitable external readout integrated circuits (ROIC) and transformed into an electric signal.

- LEDs which are reverse-biased to act as photodiodes. See LEDs as Photodiode Light Sensors.
- Photoresistors or *Light Dependent Resistors* (LDR) which change resistance according to light intensity. Normally the resistance of LDRs decreases with increasing intensity of light falling on it.
- Photodiodes which can operate in photovoltaic mode or photoconductive mode.
- Phototransistors, which act like amplifying photodiodes.
- Quantum dot photoconductors or photodiodes, which can handle wavelengths in the visible and infrared spectral regions.
- Semiconductor detectors are employed in gamma and X-ray spectrometry and as particle detectors.
- Silicon drift detectors (SDDs) are X-ray radiation detectors used in x-ray spectrometry (EDS) and electron microscopy (EDX).

Photovoltaic

- Photovoltaic cells or solar cells which produce a voltage and supply an electric current when illuminated.

Thermal

- Bolometers measure the power of incident electromagnetic radiation via the heating of a material with a temperature-dependent electrical resistance. A microbolometer is a specific type of bolometer used as a detector in a thermal camera.
- Cryogenic detectors are sufficiently sensitive to measure the energy of single x-ray, visible and infrared photons.
- Pyroelectric detectors detect photons through the heat they generate and the subsequent voltage generated in pyroelectric materials.
- Golay cells detect photons by the heat they generate in a gas-filled chamber, causing the gas to expand and deform a flexible membrane whose deflection is measured.

Photochemical

- Photoreceptor cells in the retina detect light through, for instance, a rhodopsin photon-induced chemical cascade.
- Chemical detectors, such as photographic plates, in which a silver halide molecule is split into an atom of metallic silver and a halogen atom. The photographic developer causes adjacent molecules to split similarly.

Polarization

- The photorefractive effect is used in holographic data storage.
- Polarization-sensitive photodetectors use optically anisotropic materials to detect photons of a desired linear polarization.

Graphene/Silicon Photodetectors

A graphene/n-type silicon heterojunction has been demonstrated to exhibit strong rectifying behavior and high photoresponsivity. Graphene is coupled with silicon quantum dots (Si QDs) on top of bulk Si to form a hybrid photodetector. Si QDs cause an increase of the built-in potential of the graphene/Si Schottky junction while reducing the optical reflection of the photodetector. Both the electrical and optical contributions of Si QDs enable a superior performance of the photodetector..

Frequency Range

In 2014 a technique for extending semiconductor-based photodetector's frequency range to longer, lower-energy wavelengths. Adding a light source to the device effectively "primed" the detector so that in the presence of long wavelengths, it fired on wavelengths that otherwise lacked the energy to do so.

References

- Enss, Christian (Editor) (2005). Cryogenic Particle Detection. Springer, Topics in applied physics 99. ISBN 3-540-20113-0.
- Paschotta, Dr. Rüdiger. "Encyclopedia of Laser Physics and Technology - photodetectors, photodiodes, phototransistors, pyroelectric photodetectors, array, powermeter, noise". www.rp-photonics.com. Retrieved 2016-05-31.
- Claycombe, Ann (2014-04-14). "Research finds "tunable" semiconductors will allow better detectors, solar cells". Rdmag.com. Retrieved 2014-08-24.

2

Essential Concepts of Photodetectors

Responsivity, quantum efficiency, specific detectivity and dark current are some of the significant concepts of photodetectors. Responsivity measures electrical outputs per optical input whereas the merit used to measure the performance of a photodetector is known as specific detectivity. The following text unfolds its crucial aspects in a critical yet systematic manner.

Responsivity

Responsivity measures the input–output gain of a detector system. In the specific case of a photodetector, responsivity measures the electrical output per optical input.

The responsivity of a photodetector is usually expressed in units of either amperes or volts per watt of incident radiant power. For a system that responds linearly to its input, there is a unique responsivity. For nonlinear systems, the responsivity is the local slope. Many common photodetectors respond linearly as a function of the incident power.

Responsivity is a function of the wavelength of the incident radiation and of the sensor properties, such as the bandgap of the material of which the photodetector is made. One simple expression for the responsivity R of a photodetector in which an optical signal is converted into an electric current (known as a photocurrent) is

$$R = \eta \frac{q}{hf} \approx \eta \frac{\lambda_{(\mu m)}}{1.23985(\mu m \times W / A)}$$

where η is the quantum efficiency (the conversion efficiency of photons to electrons) of the detector for a given wavelength, q is the electron charge, f is the frequency of the optical signal, and h is Planck's constant. This expression is also given in terms of λ, the wavelength of the optical signal, and has units of amperes per watt (A/W).

The term responsivity is also used to summarize input–output relationship in non-electrical systems. For example, a neuroscientist may measure how neurons in the visual pathway respond to light. In this case, responsivity summarizes the change in the neural response per unit signal strength. The responsivity in these applications can have a variety of units. The signal strength typically is controlled by varying either intensity (intensity-response function) or contrast (contrast-response function). The neural response measure depends on the part of the nervous system under study. For example, at the level of the retinal cones, the response might be in photocurrent. In the central nervous system the response is usually spikes per second. In functional neuroimaging, the response measure is usually BOLD contrast. The responsivity units reflect the relevant stimulus and physiological units.

When describing an amplifier, the more common term is gain.

Deprecated synonym **sensitivity.** A system's sensitivity is the inverse of the stimulus level required to produce a threshold response, with the threshold typically chosen just above the noise level.

Quantum Efficiency

The term quantum efficiency (QE) may apply to incident photon to converted electron (IPCE) ratio, of a photosensitive device or it may refer to the TMR effect of a Magnetic Tunnel Junction.

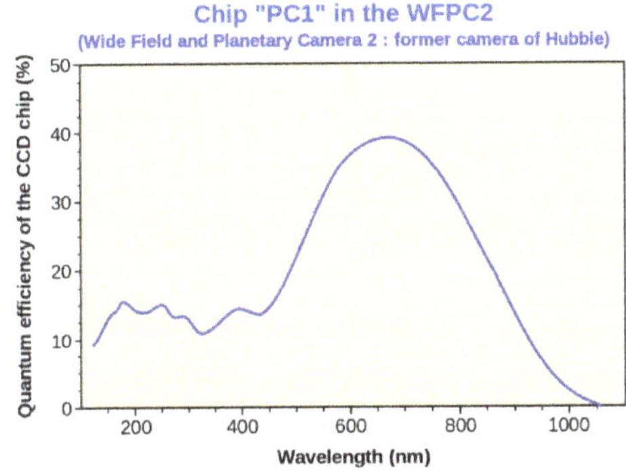

A graph showing variation of quantum efficiency with wavelength of a CCD chip in the Hubble Space Telescope's Wide Field and Planetary Camera 3.

This article deals with the term as a measurement of a device's electrical sensitivity to light. In a charge-coupled device (CCD) it is the percentage of photons hitting the device's photoreactive surface that produce charge carriers. It is measured in electrons per photon or amps per watt. Since the energy of a photon is inversely proportional to its wavelength, QE is often measured over a range of different wavelengths to characterize a device's efficiency at each photon energy level. The QE for photons with energy below the band gap is zero. Photographic film typically has a QE of much less than 10%, while CCDs can have a QE of well over 90% at some wavelengths.

Quantum Efficiency of Solar Cells

A solar cell's quantum efficiency value indicates the amount of current that the cell will produce when irradiated by photons of a particular wavelength. If the cell's quantum efficiency is integrated over the whole solar electromagnetic spectrum, one can evaluate the amount of current that the cell will produce when exposed to sunlight. The ratio between this energy-production value and the highest possible energy-production value for the cell (i.e., if the QE were 100% over the whole spectrum) gives the cell's overall energy conversion efficiency value. Note that in the event of multiple exciton generation (MEG), quantum efficiencies of greater than 100% may be achieved since the incident photons have more than twice the band gap energy and can create two or more electron-hole pairs per incident photon.

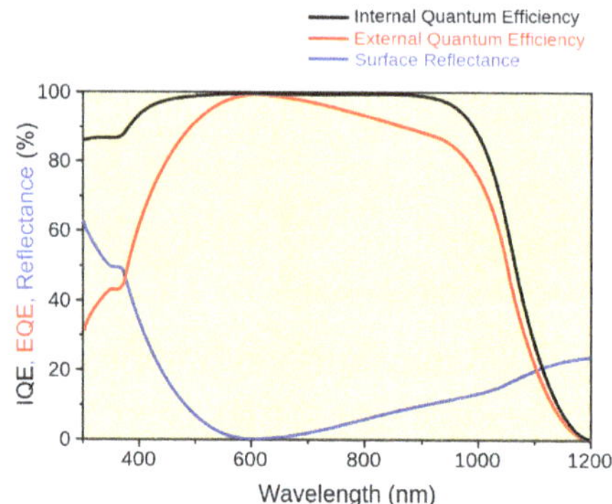

A graph showing variation of internal quantum efficiency, external quantum efficiency, and reflectance with wavelength of a crystalline silicon solar cell.

Types of Quantum Efficiency

Two types of quantum efficiency of a solar cell are often considered:

- External Quantum Efficiency (EQE) is the ratio of the number of charge carriers collected by the solar cell to the number of photons of a given energy *shining on the solar cell from outside* (incident photons).
- Internal Quantum Efficiency (IQE) is the ratio of the number of charge carriers collected by the solar cell to the number of photons of a given energy that shine on the solar cell from outside *and* are absorbed by the cell.

The IQE is always larger than the EQE. A low IQE indicates that the active layer of the solar cell is unable to make good use of the photons. To measure the IQE, one first measures the EQE of the solar device, then measures its transmission and reflection, and combines these data to infer the IQE.

$$\text{EQE} = \frac{\text{electrons/sec}}{\text{photons/sec}} = \frac{\text{current} / (\text{charge of one electron})}{(\text{total power of photons}) / (\text{energy of one photon})}$$

$$\text{IQE} = \frac{\text{electrons/sec}}{\text{absorbed photons/sec}} = \frac{\text{EQE}}{\text{1-Reflection-Transmission}}$$

The external quantum efficiency therefore depends on both the absorption of light and the collection of charges. Once a photon has been absorbed and has generated an electron-hole pair, these charges must be separated and collected at the junction. A "good" material avoids charge recombination. Charge recombination causes a drop in the external quantum efficiency.

The ideal quantum efficiency graph has a square shape, where the QE value is fairly constant across the entire spectrum of wavelengths measured. However, the QE for most solar cells is reduced because of the effects of recombination, where charge carriers are not able to move into an external circuit. The same mechanisms that affect the collection probability also affect the QE. For example, modifying the front surface can affect carriers generated near the surface. And because high-energy (blue) light is absorbed very close to the surface, considerable recombination at the front surface will affect the "blue" portion of the QE. Similarly, lower energy (green) light is absorbed in the bulk of a solar cell, and a low diffusion length will affect the collection probability from the solar cell bulk, reducing the QE in the green portion of the spectrum. Generally, solar cells on the market today do not produce much electricity from ultraviolet and infrared light (<400 nm and >1100 nm wavelengths, respectively); these wavelengths of light are either filtered out or are absorbed by the cell, thus heating the cell. That heat is wasted energy, and could damage the cell.

Quantum efficiency of Image Sensors : Quantum efficiency (QE) is the fraction of photon flux that contributes to the photocurrent in a photodetector or a pixel. Quantum efficiency is one of the most important parameters used to evaluate the quality of a detector and is often called the spectral response to reflect its wavelength dependence. It is defined as the number of signal electrons created per incident photon. In some cases it can exceed 100% (i.e. when more than one electron is created per incident photon).

EQE mapping : Conventional measurement of the EQE will give the efficiency of the overall device. However it is often useful to have a map of the EQE over large area of the device. This mapping provides an efficient way to visualize the homogeneity and/or the defects in the sample. It was realized by researchers from the Institute of Researcher and Development on Photovoltaic Energy (IRDEP) who calculated the EQE mapping from electroluminescence measurements taken with an hyperspectral imager.

Spectral Responsivity

Spectral responsivity is a similar measurement, but it has different units: amperes per watt (A/W); (i.e. how much current comes out of the device per incoming photon of a given energy and wavelength). Both the quantum efficiency and the responsivity are functions of the photons' wavelength (indicated by the subscript λ).

To convert from responsivity (R_λ, in A/W) to QE_λ (on a scale 0 to 1):

$$QE_\lambda = \frac{R_\lambda}{\lambda} \times \frac{hc}{e} \approx \frac{R_\lambda}{\lambda} \times (1240\ \mathrm{W \cdot nm/A})$$

where λ is the wavelength in nm, h is the Planck constant, c is the speed of light in a vacuum, and e is the elementary charge.

Determination

$$QE_\lambda = \eta = \frac{N_e}{N_v}$$

where N_e = number of electrons produced, N_v = number of photons absorbed.

$$\frac{N_v}{t} = \Phi_o \frac{\lambda}{hc}$$

Assuming each photon absorbed in the depletion layer produces a viable electron-hole pair, and all other photons do not,

$$\frac{N_e}{t} = \Phi_\xi \frac{\lambda}{hc}$$

where t is the measurement time (in seconds), Φ_o = incident optical power in watts, Φ_ξ = optical power absorbed in depletion layer, also in watts.

Specific Detectivity

Specific detectivity, or D^*, for a photodetector is a figure of merit used to characterize performance, equal to the reciprocal of noise-equivalent power (NEP), normalized per square root of the sensor's area and frequency bandwidth (reciprocal of twice the integration time).

Specific detectivity is given by $\frac{\sqrt{Af}}{NEP}$, where A is the area of the photosensitive region of the detector and f is the frequency bandwidth. It is commonly expressed in *Jones* units ($cm \cdot \sqrt{Hz} / W$) in honor of Robert Clark Jones who originally defined it.

Given that noise-equivalent power can be expressed as a function of the responsivity $\Re$ (in units of A/W or $V >$) and the noise spectral density S_n (in units of $A/Hz^{1/2}$ or $V/Hz^{1/2}$) as $NEP = \frac{S_n}{\Re}$, it's common to see the specific detectivity expressed as $D^* = \frac{\Re \cdot \sqrt{A}}{S_n}$.

It is often useful to express the specific detectivity in terms of relative noise levels present in the device. A common expression is given below.

$$D^* = \frac{q\lambda\eta}{hc}\left[\frac{4kT}{R_0 A} + 2q^2\eta\Phi_b\right]^{-1/2}$$

With q as the electronic charge, λ is the wavelength of interest, h is Planck's constant, c is the speed of light, k is Boltzmann's constant, T is the temperature of the detector, R_0A is the zero-bias dynamic resistance area product (often measured experimentally, but also expressible in noise level assumptions), η is the quantum efficiency of the device, and Φ_b is the total flux of the source (often a blackbody) in photons/sec/cm².

Detectivity Measurement

Detectivity can be measured from a suitable optical setup using known parameters. You will need a known light source with known irradiance at a given standoff distance. The incoming light source will be chopped at a certain frequency, and then each wavelet will be integrated over a given time constant over a given number of frames.

In detail, we compute the bandwidth Δf directly from the integration time constant t_c.

$$\Delta f = \frac{1}{2t_c}$$

Next, an rms signal and noise needs to be measured from a set of N frames. This is done either directly by the instrument, or done as post-processing.

$$Signal_{rms} = \sqrt{\frac{1}{N}\left(\sum_i^N Signal_i^2\right)}$$

$$Noise_{rms} = \sigma^2 = \sqrt{\frac{1}{N}\sum_i^N (Signal_i - Signal_{avg})^2}$$

Now, the computation of the radiance H in W/sr/cm² must be computed where cm² is the emitting area. Next, emitting area must be converted into a projected area and the solid angle; this product is often called the etendue. This step can be obviated by the use of a calibrated source, where the exact number of photons/s/cm² is known at the detector. If this is unknown, it can be estimated using the black-body radiation equation, detector active area A_d and the etendue. This ultimately converts the outgoing radiance of the black body in W/sr/cm² of emitting area into one of W observed on the detector.

The broad-band responsivity, is then just the signal weighted by this wattage.

$$R = \frac{Signal_{rms}}{HG} = \frac{Signal}{\int dH dA_d d\Omega_{BB}}$$

Where,

- R is the responsivity in units of Signal / W, (or sometimes V/W or A/W)

- H is the outgoing radiance from the black body (or light source) in W/sr/cm² of emitting area
- G is the total integrated etendue between the emitting source and detector surface
- A_d is the detector area
- Ω_{BB} is the solid angle of the source projected along the line connecting it to the detector surface.

From this metric noise-equivalent power can be computed by taking the noise level over the responsivity.

$$NEP = \frac{Noise_{rms}}{R} = \frac{Noise_{rms}}{Signal_{rms}} HG$$

Similarly, noise-equivalent irradiance can be computed using the responsivity in units of photons/s/W instead of in units of the signal. Now, the detectivity is simply the noise-equivalent power normalized to the bandwidth and detector area.

$$D^* = \frac{\sqrt{\Delta f A_d}}{NEP} = \frac{\sqrt{\Delta f A_d}}{HG} \frac{Signal_{rms}}{Noise_{rms}}$$

Dark Current (Physics)

In physics and in electronic engineering, dark current is the relatively small electric current that flows through photosensitive devices such as a photomultiplier tube, photodiode, or charge-coupled device even when no photons are entering the device; it consists of the charges generated in the detector when no outside radiation is entering the detector. It is referred to as reverse bias leakage current in non-optical devices and is present in all diodes. Physically, dark current is due to the random generation of electrons and holes within the depletion region of the device.

The charge generation rate is related to specific crystallographic defects within the depletion region. Dark-current spectroscopy can be used to determine the defects present by monitoring the peaks in the dark current histogram's evolution with temperature.

Dark current is one of the main sources for noise in image sensors such as charge-coupled devices. The pattern of different dark currents can result in a fixed-pattern noise; dark frame subtraction can remove an estimate of the mean fixed pattern, but there still remains a temporal noise, because the dark current itself has a shot noise.

Measuring Dark Current

In 2016, Belgium-based technology company Xenics developed an ultra-low-light level detector system using InGaAs sensors. To determine its dark current characteristics, tests were performed using targets at 300°K and then at 80°K. They determined that the sensors' dark current fell in a tight range of 19 to 26 electrons/pixel per second.

References

- Träger, Frank (2012). Handbook of Lasers and Optics. Berlin Heidelberg: Springer. pp. 601, 603. ISBN 9783642194092.
- Kenneth W. Busch, Marianna A. Busch (1990). Multielement Detection Systems for Spectrochemical Analysis. Wiley-Interscience. ISBN 0-471-81974-3.
- Dark Current and Influence of Target Emissivity, Photonics & Imaging Technology, September 2016, pp. 11-14.
- A. Delamarre; et al. (2014). "Quantitative luminescence mapping of $Cu(In,Ga)Se_2$ thin-film solar cells". Progress in Photovoltaics. doi:10.1002/pip.2555.
- Delamarre; et al. (2013). "Evaluation of micrometer scale lateral fluctuations of transport properties in CIGS solar cells". Proc. Of SPIE. 100. doi:10.1117/12.2004323.

3

Photodiode: An Overview

A photodiode helps in the conversion of light into electric current. Some of the applications of photodiodes are photoconductors, charge coupled devices and photomultiplier tubes. Common devices that have photodiodes are smoke detectors, remote controllers used to control televisions and air conditioner.

Photodiode

A photodiode is a semiconductor device that converts light into current. The current is generated when photons are absorbed in the photodiode. A small amount of current is also produced when no light is present. Photodiodes may contain optical filters, built-in lenses, and may have large or small surface areas. Photodiodes usually have a slower response time as their surface area increases. The common, traditional solar cell used to generate electric solar power is a large area photodiode.

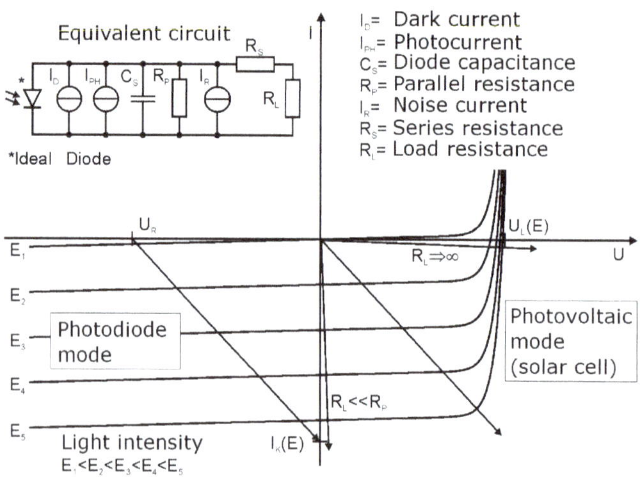

I-V characteristic of a photodiode. The linear load lines represent the response of the external circuit: I=(Applied bias voltage-Diode voltage)/Total resistance. The points of intersection with the curves represent the actual current and voltage for a given bias, resistance and illumination.

Photodiodes are similar to regular semiconductor diodes except that they may be either exposed (to detect vacuum UV or X-rays) or packaged with a window or optical fiber connection to allow light to reach the sensitive part of the device. Many diodes designed for use specifically as a photodiode use a PIN junction rather than a p–n junction, to increase the speed of response. A photodiode is designed to operate in reverse bias.

Principle of Operation

A photodiode is a p–n junction or PIN structure. When a photon of sufficient energy strikes the diode, it creates an electron-hole pair. This mechanism is also known as the inner photoelectric effect. If the absorption occurs in the junction's depletion region, or one diffusion length away from it, these carriers are swept from the junction by the built-in electric field of the depletion region. Thus holes move toward the anode, and electrons toward the cathode, and a photocurrent is produced. The total current through the photodiode is the sum of the dark current (current that is generated in the absence of light) and the photocurrent, so the dark current must be minimized to maximize the sensitivity of the device.

Photovoltaic Mode

When used in zero bias or *photovoltaic mode*, the flow of photocurrent out of the device is restricted and a voltage builds up. This mode exploits the photovoltaic effect, which is the basis for solar cells – a traditional solar cell is just a large area photodiode.

Photoconductive Mode

In this mode the diode is often reverse biased (with the cathode driven positive with respect to the anode). This reduces the response time because the additional reverse bias increases the width of the depletion layer, which decreases the junction's capacitance. The reverse bias also increases the dark current without much change in the photocurrent. For a given spectral distribution, the photocurrent is linearly proportional to the illuminance (and to the irradiance).

Although this mode is faster, the photoconductive mode tends to exhibit more electronic noise. The leakage current of a good PIN diode is so low (<1 nA) that the Johnson–Nyquist noise of the load resistance in a typical circuit often dominates.

Other Modes of Operation

Avalanche photodiodes are photodiodes with structure optimized for operating with high reverse bias, approaching the reverse breakdown voltage. This allows each *photo-generated* carrier to be multiplied by avalanche breakdown, resulting in internal gain within the photodiode, which increases the effective *responsivity* of the device.

A phototransistor is a light-sensitive transistor. A common type of phototransistor, called a photobipolar transistor, is in essence a bipolar transistor encased in a transparent case so that light can reach the *base–collector junction*. It was invented by Dr. John N. Shive (more famous for his wave machine) at Bell Labs in 1948, but it was not announced until 1950. The electrons that are generated by photons in the base–collector junction are injected into the base, and this photodiode current is amplified by the transistor's current gain β (or h_{fe}). If the base and collector leads are used and the emitter is left unconnected, the phototransistor becomes a photodiode. While phototransistors have a higher responsivity for light they are not able to detect low levels of light any better than photodiodes. Phototransistors also have significantly longer response times. Field-effect phototransistors, also known as photoFETs, are light-sensitive field-effect transistors. Unlike photobipolar transistors, photoFETs control drain-source current by creating a gate voltage.

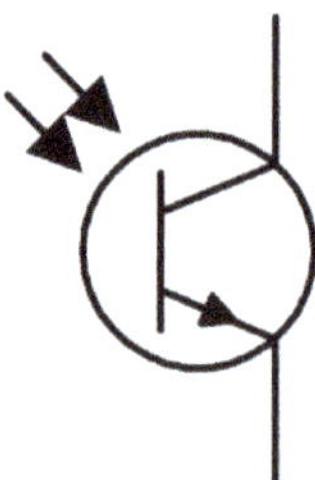

Electronic symbol for a phototransistor

Materials

The material used to make a photodiode is critical to defining its properties, because only photons with sufficient energy to excite electrons across the material's bandgap will produce significant photocurrents.

Materials commonly used to produce photodiodes include:

Material	Electromagnetic spectrum wavelength range (nm)
Silicon	190–1100
Germanium	400–1700
Indium gallium arsenide	800–2600
Lead(II) sulfide	<1000–3500
Mercury cadmium telluride	400–14000

Because of their greater bandgap, silicon-based photodiodes generate less noise than germanium-based photodiodes.

Unwanted Photodiode Effects

Any p–n junction, if illuminated, is potentially a photodiode. Semiconductor devices such as transistors and ICs contain p–n junctions, and will not function correctly if they are illuminated by unwanted electromagnetic radiation (light) of wavelength suitable to produce a photocurrent; this is avoided by encapsulating devices in opaque housings. If these housings are not completely opaque to high-energy radiation (ultraviolet, X-rays, gamma rays), transistors and ICs can malfunction due to induced photo-currents. Background radiation from the packaging is also significant. Radiation hardening mitigates these effects.

The infamous Raspberry Pi 2 xenon flash hanging bug, whereby the unit crashes when a flash photo is taken of it, is a result of the flash's intense ultraviolet emissions disrupting a switch mode power supply controller chip in bare-die wafer-scale package, with the resulting power surge causing the main processor to lock up.

Features

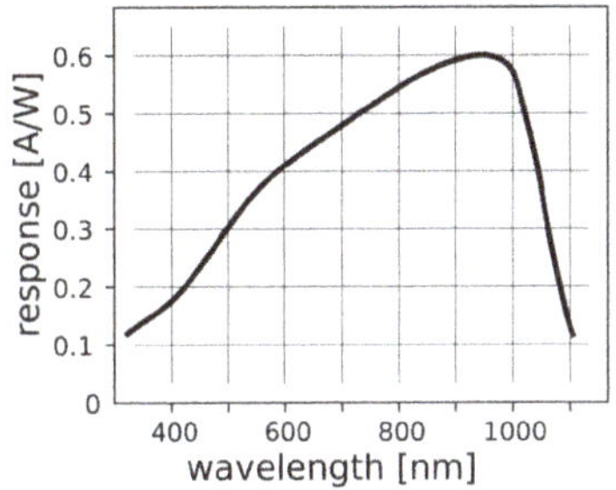

Response of a silicon photo diode vs wavelength of the incident light

Critical performance parameters of a photodiode include:

Responsivity

> The Spectral responsivity is a ratio of the generated photocurrent to incident light power, expressed in A/W when used in photoconductive mode. The wavelength-dependence may also be expressed as a *Quantum efficiency*, or the ratio of the number of photogenerated carriers to incident photons, a unitless quantity.

Dark current

> The current through the photodiode in the absence of light, when it is operated in photoconductive mode. The dark current includes photocurrent generated by background radiation and the saturation current of the semiconductor junction. Dark current must be accounted for by calibration if a photodiode is used to make an accurate optical power measurement, and it is also a source of noise when a photodiode is used in an optical communication system.

Response time

> A photon absorbed by the semiconducting material will generate an electron-hole pair which will in turn start moving in the material under the effect of the electric field and thus generate a current. The finite duration of this current is known as the transit-time spread and can be evaluated by using Ramo's theorem. One can also show with this theorem that the total charge generated in the external circuit is well e and not 2e as might seem by the presence of the two carriers. Indeed, the integral of the current due to both electron and hole over time must be equal to e. The resistance and capacitance of the photodiode and

the external circuitry give rise to another response time known as RC time constant $\tau = RC$. This combination of R and C integrates the photoresponse over time and thus lengthens the impulse response of the photodiode. When used in an optical communication system, the response time determines the bandwidth available for signal modulation and thus data transmission.

Noise-equivalent power

(NEP) The minimum input optical power to generate photocurrent, equal to the rms noise current in a 1 hertz bandwidth. NEP is essentially the minimum detectable power. The related characteristic detectivity (D) is the inverse of NEP, 1/NEP. There is also the specific detectivity (D^*) which is the detectivity multiplied by the square root of the area (A) of the photodetector, ($D^* = D\sqrt{A}$) for a 1 Hz bandwidth. The specific detectivity allows different systems to be compared independent of sensor area and system bandwidth; a higher detectivity value indicates a low-noise device or system. Although it is traditional to give (D^*) in many catalogues as a measure of the diode's quality, in practice, it is hardly ever the key parameter.

When a photodiode is used in an optical communication system, all these parameters contribute to the *sensitivity* of the optical receiver, which is the minimum input power required for the receiver to achieve a specified *bit error rate*.

Applications

P–n photodiodes are used in similar applications to other photodetectors, such as photoconductors, charge-coupled devices, and photomultiplier tubes. They may be used to generate an output which is dependent upon the illumination (analog; for measurement and the like), or to change the state of circuitry (digital; either for control and switching, or digital signal processing).

Photodiodes are used in consumer electronics devices such as compact disc players, smoke detectors, and the receivers for infrared remote control devices used to control equipment from televisions to air conditioners. For many applications either photodiodes or photoconductors may be used. Either type of photosensor may be used for light measurement, as in camera light meters, or to respond to light levels, as in switching on street lighting after dark.

Photosensors of all types may be used to respond to incident light, or to a source of light which is part of the same circuit or system. A photodiode is often combined into a single component with an emitter of light, usually a light-emitting diode (LED), either to detect the presence of a mechanical obstruction to the beam (slotted optical switch), or to couple two digital or analog circuits while maintaining extremely high electrical isolation between them, often for safety (optocoupler). The combination of LED and photodiode is also used in many sensor systems to characterize different types of products based on their optical absorbance.

Photodiodes are often used for accurate measurement of light intensity in science and industry. They generally have a more linear response than photoconductors.

They are also widely used in various medical applications, such as detectors for computed tomography (coupled with scintillators), instruments to analyze samples (immunoassay), and pulse oximeters.

PIN diodes are much faster and more sensitive than p–n junction diodes, and hence are often used for optical communications and in lighting regulation.

P–n photodiodes are not used to measure extremely low light intensities. Instead, if high sensitivity is needed, avalanche photodiodes, intensified charge-coupled devices or photomultiplier tubes are used for applications such as astronomy, spectroscopy, night vision equipment and laser rangefinding.

Pinned photodiode is not a PIN photodiode, it has p+/n/p regions in it. It has a shallow P+ implant in N type diffusion layer over a P-type epitaxial substrate layer. It is used in CMOS Active pixel sensor.

Comparison with Photomultipliers

Advantages compared to photomultipliers:

1. Excellent linearity of output current as a function of incident light
2. Spectral response from 190 nm to 1100 nm (silicon), longer wavelengths with other semiconductor materials
3. Low noise
4. Ruggedized to mechanical stress
5. Low cost
6. Compact and light weight
7. Long lifetime
8. High quantum efficiency, typically 60–80%
9. No high voltage required

Disadvantages compared to photomultipliers:

1. Small area
2. No internal gain (except avalanche photodiodes, but their gain is typically 10^2–10^3 compared to 10^5-10^8 for the photomultiplier)

3. Much lower overall sensitivity
4. Photon counting only possible with specially designed, usually cooled photodiodes, with special electronic circuits
5. Response time for many designs is slower
6. latent effect

Photodiode Array

A one-dimensional array of hundreds or thousands of photodiodes can be used as a position sensor, for example as part of an angle sensor. One advantage of photodiode arrays (PDAs) is that they allow for high speed parallel read out since the driving electronics may not be built in like a traditional CMOS or CCD sensor.

A 2 x 2 cm photodiode array chip with more than 200 diodes

Avalanche Photodiode

An avalanche photodiode (APD) is a highly sensitive semiconductor electronic device that exploits the photoelectric effect to convert light to electricity. APDs can be thought of as photodetectors that provide a built-in first stage of gain through avalanche multiplication. From a functional standpoint, they can be regarded as the semiconductor analog to photomultipliers. By applying a high reverse bias voltage (typically 100-200 V in silicon), APDs show an internal current gain effect (around 100) due to impact ionization (avalanche effect). However, some silicon APDs employ alternative doping and beveling techniques compared to traditional APDs that allow greater voltage to be applied (> 1500 V) before breakdown is reached and hence a greater operating gain (> 1000). In general, the higher the reverse voltage the higher the gain. Among the various expressions for the APD multiplication factor (*M*), an instructive expression is given by the formula

Avalanche photodiode.

$$M = \frac{1}{1 - \int_0^L \alpha(x)dx}$$

where L is the space charge boundary for electrons and α is the multiplication coefficient for electrons (and holes). This coefficient has a strong dependence on the applied electric field strength, temperature, and doping profile. Since APD gain varies strongly with the applied reverse bias and temperature, it is necessary to control the reverse voltage to keep a stable gain. Avalanche photodiodes therefore are more sensitive compared to other semiconductor photodiodes.

If very high gain is needed (10^5 to 10^6), certain APDs (single-photon avalanche diodes) can be operated with a reverse voltage above the APD's breakdown voltage. In this case, the APD needs to have its signal current limited and quickly diminished. Active and passive current quenching techniques have been used for this purpose. APDs that operate in this high-gain regime are in Geiger mode. This mode is particularly useful for single photon detection provided that the dark count event rate is sufficiently low.

A typical application for APDs is laser rangefinders and long range fiber optic telecommunication. New applications include positron emission tomography and particle physics. APD arrays are becoming commercially available.

APD applicability and usefulness depends on many parameters. Two of the larger factors are: quantum efficiency, which indicates how well incident optical photons are absorbed and then used to generate primary charge carriers; and total leakage current, which is the sum of the dark current and photocurrent and noise. Electronic dark noise components are series and parallel noise. Series noise, which is the effect of shot noise, is basically proportional to the APD capacitance while the parallel noise is associated with the fluctuations of the APD bulk and surface dark currents. Another noise source is the excess noise factor, ENF. It describes the statistical noise that is inherent with the stochastic APD multiplication process.

Materials

In principle any semiconductor material can be used as a multiplication region:

- Silicon will detect in the visible and near infrared, with low multiplication noise (excess noise).
- Germanium (Ge) will detect infrared out to a wavelength of 1.7 µm, but has high multiplication noise.
- InGaAs will detect out to longer than 1.6 µm, and has less multiplication noise than Ge. It is normally used as the absorption region of a heterostructure diode, most typically involving InP as a substrate and as a multiplication layer. This material system is compatible with an absorption window of roughly 0.9-1.7 µm. InGaAs exhibits a high absorption coefficient at the wavelengths appropriate to high-speed telecommunications using optical fibers, so only a few micrometres of InGaAs are required for nearly 100% light absorption. The excess noise factor is low enough to permit a gain-bandwidth product in excess of 100 GHz for a simple InP/InGaAs system, and up to 400 GHz for InGaAs on silicon. Therefore, high speed operation is possible: commercial devices are available to speeds of at least 10 Gbit/s.
- Gallium nitride based diodes have been used for operation with ultraviolet light.
- HgCdTe based diodes operate in the infrared, typically out to a maximum wavelength of about 14 µm, but require cooling to reduce dark currents. Very low excess noise can be achieved in this material system.

Excess Noise

Excess noise refers to the noise due to the multiplication process at a gain, M is denoted by ENF(M) and can often be expressed as:

$$\mathrm{ENF} = \kappa M + \left(2 - \frac{1}{M}\right)(1-\kappa)$$

where κ is the ratio of the hole impact ionization rate to that of electrons. For an electron multiplication device it is given by the hole impact ionization rate divided by the electron impact ionization rate. It is desirable to have a large asymmetry between these rates to minimize ENF(M), since ENF(M) is one of the main factors that limit, among other things, the best possible energy resolution obtainable.

Performance Limits

In addition to excess noise, there are limits to device performance associated with the capacitance, transit times and avalanche multiplication time. The capacitance increases with increasing device area and decreasing thickness. The transit times (both electrons and holes) increase with increasing thickness, implying a tradeoff between capacitance and transit time for performance. The avalanche multiplication time times the gain is given to first order by the gain-bandwidth product, which is a function of the device structure and most especially κ .

Diode

In electronics, a diode is a two-terminal electronic component that conducts primarily in one direction (asymmetric conductance); it has low (ideally zero) resistance to the flow of current in one direction, and high (ideally infinite) resistance in the other. A semiconductor diode, the most common type today, is a crystalline piece of semiconductor material with a p–n junction connected to two electrical terminals. A vacuum tube diode has two electrodes, a plate (anode) and a heated cathode. Semiconductor diodes were the first semiconductor electronic devices. The discovery of crystals' rectifying abilities was made by German physicist Ferdinand Braun in 1874. The first semiconductor diodes, called cat's whisker diodes, developed around 1906, were made of mineral crystals such as galena. Today, most diodes are made of silicon, but other semiconductors such as selenium or germanium are sometimes used.

Closeup of a diode, showing the square-shaped semiconductor crystal *(black object on left)*.

Extreme macro photo of a Chinese diode of the seventies.

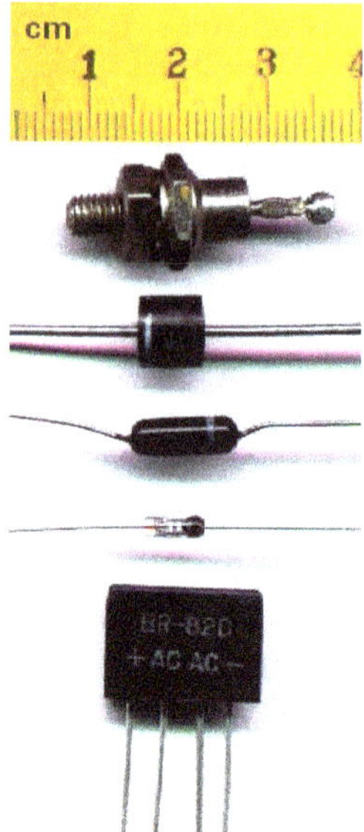

Various semiconductor diodes. Bottom: A bridge rectifier. In most diodes, a white or black painted band identifies the cathode into which electrons will flow when the diode is conducting. Electron flow is the reverse of conventional current flow.

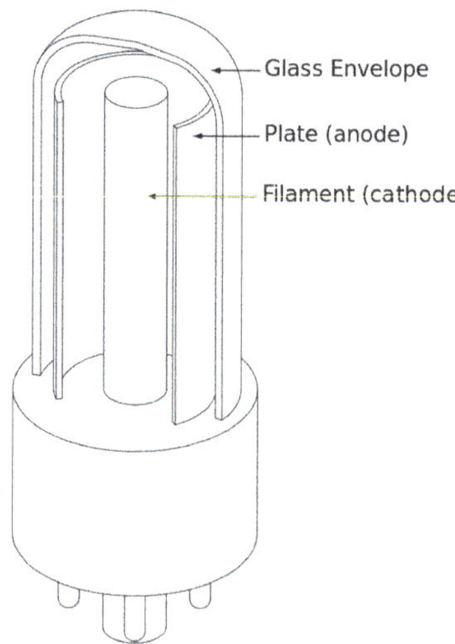

Structure of a vacuum tube diode. The filament may be bare, or more commonly (as shown here), embedded within and insulated from an enclosing cathode.

Main Functions

The most common function of a diode is to allow an electric current to pass in one direction (called the diode's *forward* direction), while blocking current in the opposite direction (the *reverse* direction). Thus, the diode can be viewed as an electronic version of a check valve. This unidirectional behavior is called rectification, and is used to convert alternating current (AC) to direct current (DC), including extraction of modulation from radio signals in radio receivers—these diodes are forms of rectifiers.

However, diodes can have more complicated behavior than this simple on–off action, because of their nonlinear current-voltage characteristics. Semiconductor diodes begin conducting electricity only if a certain threshold voltage or cut-in voltage is present in the forward direction (a state in which the diode is said to be *forward-biased*). The voltage drop across a forward-biased diode varies only a little with the current, and is a function of temperature; this effect can be used as a temperature sensor or as a voltage reference.

A semiconductor diode's current–voltage characteristic can be tailored by selecting the semiconductor materials and the doping impurities introduced into the materials during manufacture. These techniques are used to create special-purpose diodes that perform many different functions. For example, diodes are used to regulate voltage (Zener diodes), to protect circuits from high voltage surges (avalanche diodes), to electronically tune radio and TV receivers (varactor diodes), to generate radio-frequency oscillations (tunnel diodes, Gunn diodes, IMPATT diodes), and to produce light (light-emitting diodes). Tunnel, Gunn and IMPATT diodes exhibit negative resistance, which is useful in microwave and switching circuits.

Diodes, both vacuum and semiconductor, can be used as shot-noise generators.

History

Thermionic (vacuum tube) diodes and solid state (semiconductor) diodes were developed separately, at approximately the same time, in the early 1900s, as radio receiver detectors. Until the 1950s vacuum tube diodes were used more frequently in radios because the early point-contact type semiconductor diodes were less stable. In addition, most receiving sets had vacuum tubes for amplification that could easily have the thermionic diodes included in the tube (for example the 12SQ7 double diode triode), and vacuum tube rectifiers and gas-filled rectifiers were capable of handling some high voltage/high current rectification tasks better than the semiconductor diodes (such as selenium rectifiers) which were available at that time.

Vacuum Tube Diodes

In 1873, Frederick Guthrie discovered the basic principle of operation of thermionic diodes. Guthrie discovered that a positively charged electroscope could be discharged by bringing a grounded piece of white-hot metal close to it (but not actually touching it). The same did not apply to a negatively charged electroscope, indicating that the current flow was only possible in one direction.

Thomas Edison independently rediscovered the principle on February 13, 1880. At the time, Edison was investigating why the filaments of his carbon-filament light bulbs nearly always burned out at the positive-connected end. He had a special bulb made with a metal plate sealed into the glass envelope. Using this device, he confirmed that an invisible current flowed from the glowing filament through the vacuum to the metal plate, but only when the plate was connected to the positive supply.

Edison devised a circuit where his modified light bulb effectively replaced the resistor in a DC voltmeter. Edison was awarded a patent for this invention in 1884. Since there was no apparent practical use for such a device at the time, the patent application was most likely simply a precaution in case someone else did find a use for the so-called Edison effect.

About 20 years later, John Ambrose Fleming (scientific adviser to the Marconi Company and former Edison employee) realized that the Edison effect could be used as a precision radio detector. Fleming patented the first true thermionic diode, the Fleming valve, in Britain on November 16, 1904 (followed by U.S. Patent 803,684 in November 1905).

Solid-state Diodes

In 1874 German scientist Karl Ferdinand Braun discovered the "unilateral conduction" of crystals. Braun patented the crystal rectifier in 1899. Copper oxide and selenium rectifiers were developed for power applications in the 1930s.

Indian scientist Jagadish Chandra Bose was the first to use a crystal for detecting radio waves in 1894. The crystal detector was developed into a practical device for wireless telegraphy by Greenleaf Whittier Pickard, who invented a silicon crystal detector in 1903 and received a patent for it on November 20, 1906. Other experimenters tried a variety of other substances, of which the most widely used was the mineral galena (lead sulfide). Other substances offered slightly better performance, but galena was most widely used because it had the advantage of being cheap and easy to obtain. The crystal detector in these early crystal radio sets consisted of an adjustable wire point-contact, often made of gold or platinum because of their incorrodible nature (the so-called "cat's whisker"), which could be manually moved over the face of the crystal in search of a portion of that mineral with rectifying qualties. This troublesome device was superseded by thermionic diodes (vacuum tubes) by the 1920s, but after high purity semiconductor materials became available, the crystal detector returned to dominant use with the advent, in the 1950s, of inexpensive fixed-germanium diodes. Bell Labs also developed a germanium diode for microwave reception, and AT&T used these in their microwave towers that criss-crossed the nation starting in the late 1940s, carrying telephone and network television signals. Bell Labs did not develop a satisfactory thermionic diode for microwave reception.

Etymology

At the time of their invention, such devices were known as rectifiers. In 1919, the year tetrodes were invented, William Henry Eccles coined the term *diode* from the Greek roots *di* (from *δί*), meaning 'two', and *ode* (from □*δός*), meaning 'path'. (However, the word *diode* itself, as well as *triode, tetrode, pentode, hexode,* were already in use as terms of multiplex telegraphy; see, for example, *The telegraphic journal and electrical review*, September 10, 1886, p. 252).

Rectifiers

Although all diodes *rectify*, the term 'rectifier' is normally reserved for higher currents and voltages than would normally be found in the rectification of lower power signals; examples include:

- Power supply rectifiers (*half-wave, full-wave, bridge*)
- Flyback diodes

World's Smallest Diode

Researchers from the University of Georgia and Ben-Gurion University of the Negev (BGU) have developed a diode made from a molecule of DNA. Professor Bingqian Xu from the College of Engineering at the University of Georgia and his team took a single DNA molecule made from 11 base pairs and connected it to an electronic circuit a few nanometers in size. When layers of coralyne were inserted between layers of DNA, the current jumped up to 15 times larger negative versus positive, which is necessary for a nano diode.

Thermionic Diodes

A thermionic diode is a thermionic-valve device (also known as a vacuum tube, tube, or valve), consisting of a sealed evacuated glass envelope containing two electrodes: a cathode heated by a filament, and a plate (anode). Early examples were fairly similar in appearance to incandescent light bulbs.

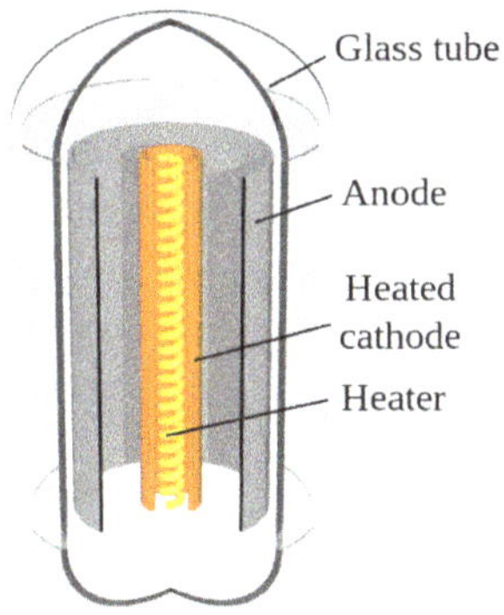

Diode vacuum tube construction

In operation, a separate current through the filament (heater), a high resistance wire made of nichrome, heats the cathode red hot (800–1000 °C), causing it to release electrons into the vacuum, a process called thermionic emission. The cathode is coated with oxides of alkaline earth metals such as barium and strontium oxides, which have a low work function, to increase the number of electrons emitted. (Some valves use *direct heating*, in which a tungsten filament acts as both heater and cathode.) The alternating voltage to be rectified is applied between the cathode and the concentric plate electrode. When the plate has a positive voltage with respect to the cathode, it electrostatically attracts the electrons from the cathode, so a current of electrons flows through the tube from cathode to plate. However, when the polarity is reversed and the plate has a negative voltage, no current flows, because the cathode electrons are not attracted to it. The unheated plate does not emit any electrons itself. So electrons can only flow through the tube in one direction, from the cathode to the anode plate.

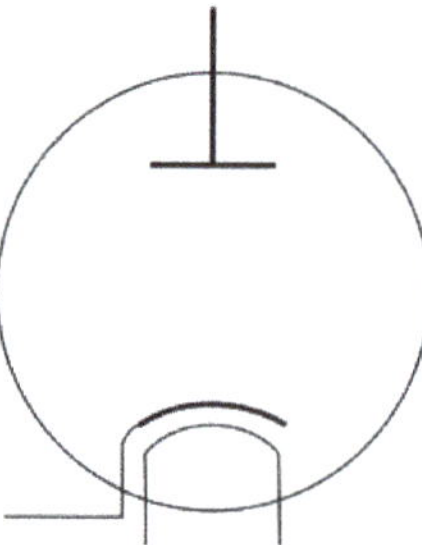

The symbol for an indirect heated vacuum-tube diode. From top to bottom, the components are the anode, the cathode, and the heater filament.

In a mercury-arc valve, an arc forms between a refractory conductive anode and a pool of liquid mercury acting as cathode. Such units were made with ratings up to hundreds of kilowatts, and were important in the development of HVDC power transmission. Some types of smaller thermionic rectifiers had mercury vapor fill to reduce their forward voltage drop and to increase current rating over thermionic hard-vacuum devices.

Throughout the vacuum tube era, valve diodes were used in analog signal applications and as rectifiers in DC power supplies in consumer electronics such as radios, televisions, and sound systems. They were replaced in power supplies beginning in the 1940s by selenium rectifiers and then by semiconductor diodes by the 1960s. Today they are still used in a few high power applications where their ability to withstand transient voltages and their robustness gives them an advantage over semiconductor devices. The recent (2012) resurgence of interest among audiophiles and recording studios in old valve audio gear such as guitar amplifiers and home audio systems has provided a market for the legacy consumer diode valves.

Semiconductor Diodes

Electronic Symbols

The symbol used for a semiconductor diode in a circuit diagram specifies the type of diode. There are alternative symbols for some types of diodes, though the differences are minor. The triangle in the symbols points to the forward direction.

Anode Cathode

Photodiode

Anode Cathode

Diode

Anode Cathode

Schottky diode

Anode Cathode

Light-emitting diode (LED)

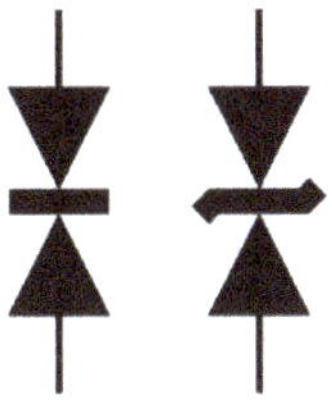

Transient-voltage-suppression diode (TVS)

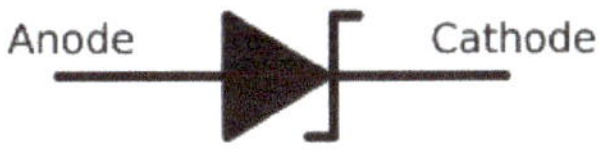

Anode Cathode

Zener diode

Anode Cathode

Varicap

Anode Cathode

Tunnel diode

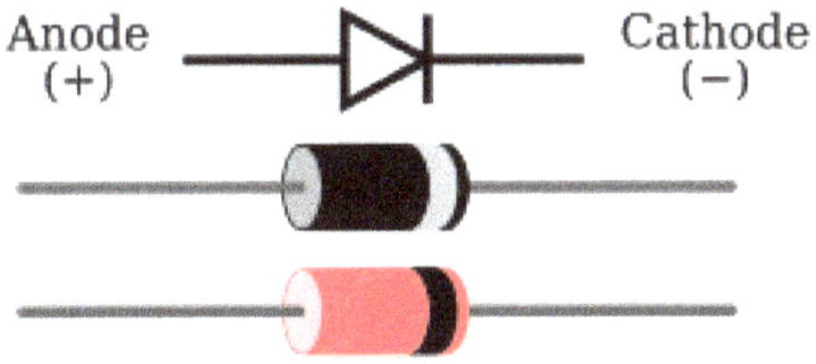

Anode (+) Cathode (−)

Typical diode packages in same alignment as diode symbol. Thin bar depicts the cathode.

A galena cat's-whisker detector, a point-contact diode.

Point-contact Diodes

A point-contact diode works the same as the junction diodes described below, but its construction is simpler. A pointed metal wire is placed in contact with an n-type semiconductor. Some metal migrates into the semiconductor to make a small p-type region around the contact. The 1N34 germanium version is still used in radio receivers as a detector and occasionally in specialized analog electronics.

Junction Diodes

p–N Junction Diode

A p–n junction diode is made of a crystal of semiconductor, usually silicon, but germanium and gallium arsenide are also used. Impurities are added to it to create a region on one side that contains negative charge carriers (electrons), called an n-type semiconductor, and a region on the other side that contains positive charge carriers (holes), called a p-type semiconductor. When the n-type and p-type materials are attached together, a momentary flow of electrons occur from the n to the p side resulting in a third region between the two where no charge carriers are present. This region is called the depletion region because there are no charge carriers (neither electrons nor holes) in it. The diode's terminals are attached to the n-type and p-regions. The boundary between these two regions, called a p–n junction, is where the action of the diode takes place. When a sufficiently higher electrical potential is applied to the P side (the anode) than to the N side (the cathode), it allows electrons to flow through the depletion region from the N-type side to the P-type side. The junction does not allow the flow of electrons in the opposite direction when the potential is applied in reverse, creating, in a sense, an electrical check valve.

Schottky Diode

Another type of junction diode, the Schottky diode, is formed from a metal–semiconductor junction rather than a p–n junction, which reduces capacitance and increases switching speed.

Current–voltage Characteristic

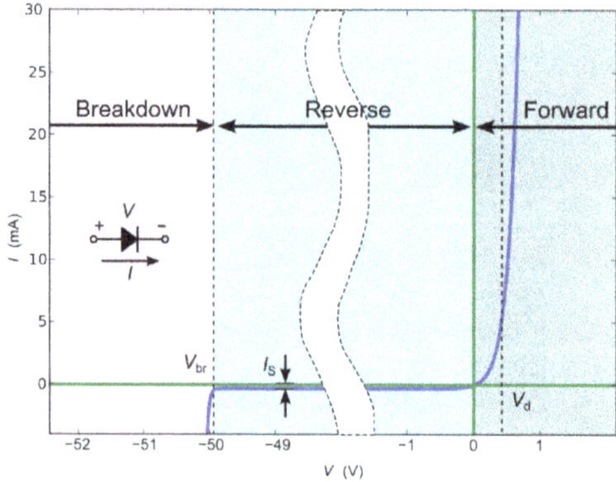

I–V (current vs. voltage) characteristics of a p–n junction diode

A semiconductor diode's behavior in a circuit is given by its current–voltage characteristic, or I–V graph (see graph below). The shape of the curve is determined by the transport of charge carriers through the so-called *depletion layer* or *depletion region* that exists at the p–n junction between differing semiconductors. When a p–n junction is first created, conduction-band (mobile) electrons from the N-doped region diffuse into the P-doped region where there is a large population of holes (vacant places for electrons) with which the electrons "recombine". When a mobile electron recombines with a hole, both hole and electron vanish, leaving behind an immobile positively charged donor (dopant) on the N side and negatively charged acceptor (dopant) on the P side. The region around the p–n junction becomes depleted of charge carriers and thus behaves as an insulator.

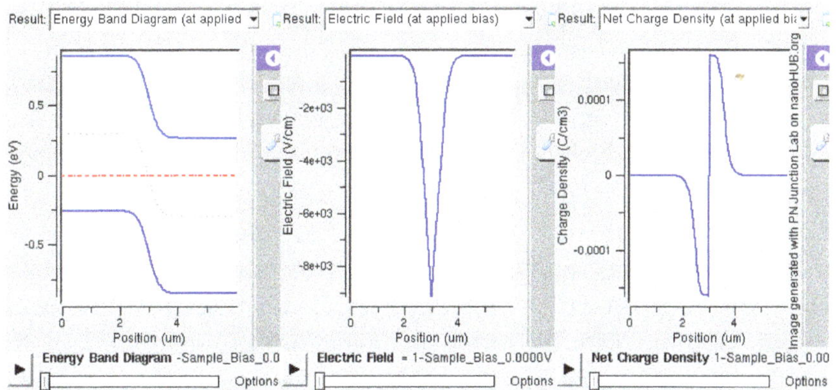

A PN junction diode in forward bias mode, the depletion width decreases. Both p and n junctions are doped at a 1e15/cm3 doping level, leading to built-in potential of ~0.59V. Observe the different Quasi Fermi levels for conduction band and valence band in n and p regions (red curves).

However, the width of the depletion region (called the depletion width) cannot grow without limit. For each electron–hole pair recombination made, a positively charged dopant ion is left behind in the N-doped region, and a negatively charged dopant ion is created in the P-doped region. As recombination proceeds and more ions are created, an increasing electric field develops through the depletion zone that acts to slow and then finally stop recombination. At this point, there is a "built-in" potential across the depletion zone.

Reverse Bias

If an external voltage is placed across the diode with the same polarity as the built-in potential, the depletion zone continues to act as an insulator, preventing any significant electric current flow (unless electron–hole pairs are actively being created in the junction by, for instance, light. This is called the *reverse bias* phenomenon.

Forward Bias

However, if the polarity of the external voltage opposes the built-in potential, recombination can once again proceed, resulting in a substantial electric current through the p–n junction (i.e. substantial numbers of electrons and holes recombine at the junction). For silicon diodes, the built-in potential is approximately 0.7 V (0.3 V for germanium and 0.2 V for Schottky). Thus, if an external voltage greater than and opposite to the built-in voltage is applied, a current will flow and the diode is said to be "turned on" as it has been given an external *forward bias*. The diode is commonly said to have a forward "threshold" voltage, above which it conducts and below which conduction stops. However, this is only an approximation as the forward characteristic is according to the Shockley equation absolutely smooth.

A diode's I–V characteristic can be approximated by four regions of operation:

1. At very large reverse bias, beyond the peak inverse voltage or PIV, a process called reverse breakdown occurs that causes a large increase in current (i.e., a large number of electrons and holes are created at, and move away from the p–n junction) that usually damages the device permanently. The avalanche diode is deliberately designed for use in that manner. In the Zener diode, the concept of PIV is not applicable. A Zener diode contains a heavily doped p–n junction allowing electrons to tunnel from the valence band of the p-type material to the conduction band of the n-type material, such that the reverse voltage is "clamped" to a known value (called the *Zener voltage*), and avalanche does not occur. Both devices, however, do have a limit to the maximum current and power they can withstand in the clamped reverse-voltage region. Also, following the end of forward conduction in any diode, there is reverse current for a short time. The device does not attain its full blocking capability until the reverse current ceases.

2. For a bias less than the PIV, the reverse current is very small. For a normal P–N rectifier diode, the reverse current through the device in the micro-ampere (μA) range is very low. However, this is temperature dependent, and at sufficiently high temperatures, a substantial amount of reverse current can be observed (mA or more).

3. With a small forward bias, where only a small forward current is conducted, the current–voltage curve is exponential in accordance with the ideal diode equation. There is a definite forward voltage at which the diode starts to conduct significantly. This is called the *knee voltage* or *cut-in voltage* and is equal to the barrier potential of the p-n junction.

4. At larger forward currents the current-voltage curve starts to be dominated by the ohmic resistance of the bulk semiconductor. The curve is no longer exponential, it is asymptotic to a straight line whose slope is the bulk resistance. This region is particularly important for power diodes. The diode can be modeled as an ideal diode in series with a fixed resistor.

In a small silicon diode operating at its rated currents, the voltage drop is about 0.6 to 0.7 volts. The value is different for other diode types—Schottky diodes can be rated as low as 0.2 V, germanium diodes 0.25 to 0.3 V, and red or blue light-emitting diodes (LEDs) can have values of 1.4 V and 4.0 V respectively.

At higher currents the forward voltage drop of the diode increases. A drop of 1 V to 1.5 V is typical at full rated current for power diodes.

Shockley Diode Equation

The *Shockley ideal diode equation* or the *diode law* (named after transistor co-inventor William Bradford Shockley) gives the I–V characteristic of an ideal diode in either forward or reverse bias (or no bias). The following equation is called the *Shockley ideal diode equation* when n, the ideality factor, is set equal to 1 :

$$I = I_S\left(e^{\frac{V_D}{nV_T}} - 1\right)$$

where

I is the diode current,

I_S is the reverse bias saturation current (or scale current),

V_D is the voltage across the diode,

V_T is the thermal voltage, and

n is the *ideality factor*, also known as the *quality factor* or sometimes *emission coefficient*. The ideality factor n typically varies from 1 to 2 (though can in some cases be higher), depending on the fabrication process and semiconductor material and is set equal to 1 for the case of an "ideal" diode (thus the n is sometimes omitted). The ideality factor was added to account for imperfect junctions as observed in real transistors. The factor mainly accounts for carrier recombination as the charge carriers cross the depletion region.

The thermal voltage V_T is approximately 25.85 mV at 300 K, a temperature close to "room temperature" commonly used in device simulation software. At any temperature it is a known constant defined by:

$$V_T = \frac{kT}{q},$$

where k is the Boltzmann constant, T is the absolute temperature of the p–n junction, and q is the magnitude of charge of an electron (the elementary charge).

The reverse saturation current, I_S, is not constant for a given device, but varies with temperature; usually more significantly than V_T, so that V_D typically decreases as T increases.

The *Shockley ideal diode equation* or the *diode law* is derived with the assumption that the only processes giving rise to the current in the diode are drift (due to electrical field), diffusion, and thermal recombination–generation (R–G) (this equation is derived by setting n = 1 above). It also assumes that the R–G current in the depletion region is insignificant. This means that the *Shockley ideal diode equation* doesn't account for the processes involved in reverse breakdown and photon-assisted R–G. Additionally, it doesn't describe the "leveling off" of the I–V curve at high forward bias due to internal resistance. Introducing the ideality factor, n, accounts for recombination and generation of carriers.

Under *reverse bias* voltages the exponential in the diode equation is negligible, and the current is a constant (negative) reverse current value of $-I_S$. The reverse *breakdown region* is not modeled by the Shockley diode equation.

For even rather small *forward bias* voltages the exponential is very large, since the thermal voltage is very small in comparison. The subtracted '1' in the diode equation is then negligible and the forward diode current can be approximated by

$$I = I_S e^{\frac{V_D}{nV_T}}$$

The use of the diode equation in circuit problems is illustrated in the article on diode modeling.

Small-signal Behavior

For circuit design, a small-signal model of the diode behavior often proves useful. A specific example of diode modeling is discussed in the article on small-signal circuits.

Reverse-recovery Effect

Following the end of forward conduction in a p–n type diode, a reverse current can flow for a short time. The device does not attain its blocking capability until the mobile charge in the junction is depleted.

The effect can be significant when switching large currents very quickly. A certain amount of "reverse recovery time" t_r (on the order of tens of nanoseconds to a few microseconds) may be required to remove the reverse recovery charge Q_r from the diode. During this recovery time, the diode can actually conduct in the reverse direction. This might give rise to a large constant current in the reverse direction for a short time while the diode is reverse biased. The magnitude of such a reverse current is determined by the operating circuit (i.e., the series resistance) and the diode is said to be in the storage-phase. In certain real-world cases it is important to consider the losses that are incurred by this non-ideal diode effect. However, when the slew rate of the current is not so severe (e.g. Line frequency) the effect can be safely ignored. For most applications, the effect is also negligible for Schottky diodes.

The reverse current ceases abruptly when the stored charge is depleted; this abrupt stop is exploited in step recovery diodes for generation of extremely short pulses.

Types of Semiconductor Diode

There are several types of p–n junction diodes, which emphasize either a different physical aspect of a diode often by geometric scaling, doping level, choosing the right electrodes, are just an application of a diode in a special circuit, or are really different devices like the Gunn and laser diode and the MOSFET:

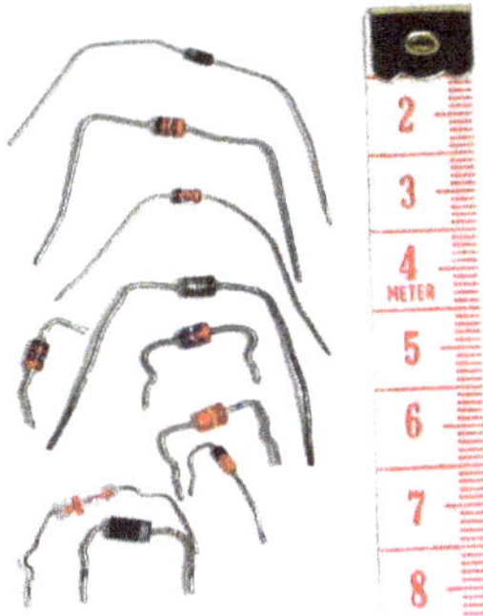

Several types of diodes. The scale is centimeters.

Normal (p–n) diodes, which operate as described above, are usually made of doped silicon or, more rarely, germanium. Before the development of silicon power rectifier diodes, cuprous oxide and later selenium was used. Their low efficiency required a much higher forward voltage to be applied (typically 1.4 to 1.7 V per "cell", with multiple cells stacked so as to increase the peak inverse voltage rating for application in high voltage rectifiers), and required a large heat sink (often an extension of the diode's metal substrate), much larger than the later silicon diode of the same current ratings would require. The vast majority of all diodes are the p–n diodes found in CMOS integrated circuits, which include two diodes per pin and many other internal diodes.

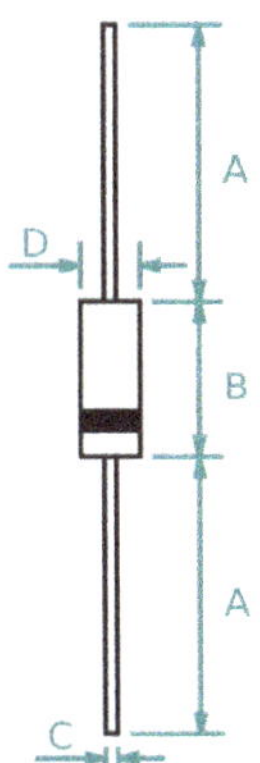

Typical datasheet drawing showing the dimensions of a DO-41 diode package

Avalanche diodes

These are diodes that conduct in the reverse direction when the reverse bias voltage exceeds the breakdown voltage. These are electrically very similar to Zener diodes (and are often mistakenly called Zener diodes), but break down by a different mechanism: the *avalanche effect*. This occurs when the reverse electric field applied across the p–n junction causes a wave of ionization, reminiscent of an avalanche, leading to a large current. Avalanche diodes are designed to break down at a well-defined reverse voltage without being destroyed. The difference between the avalanche diode (which has a reverse breakdown above about 6.2 V) and the Zener is that the channel length of the former exceeds the mean free path of the electrons, resulting in many collisions between them on the way through the channel. The only practical difference between the two types is they have temperature coefficients of opposite polarities.

Cat's whisker or crystal diodes

These are a type of point-contact diode. The cat's whisker diode consists of a thin or sharpened metal wire pressed against a semiconducting crystal, typically galena or a piece of coal. The wire forms the anode and the crystal forms the cathode. Cat's whisker diodes were also called crystal diodes and found application in the earliest radios called crystal radio receivers. Cat's whisker diodes are generally obsolete, but may be available from a few manufacturers.

Constant current diodes

These are actually JFETs with the gate shorted to the source, and function like a two-terminal current-limiting analog to the voltage-limiting Zener diode. They allow a current through them to rise to a certain value, and then level off at a specific value. Also called *CLDs, constant-current diodes, diode-connected transistors*, or *current-regulating diodes*.

Esaki or tunnel diodes

These have a region of operation showing negative resistance caused by quantum tunneling, allowing amplification of signals and very simple bistable circuits. Because of the high carrier concentration, tunnel diodes are very fast, may be used at low (mK) temperatures, high magnetic fields, and in high radiation environments. Because of these properties, they are often used in spacecraft.

Gunn diodes

These are similar to tunnel diodes in that they are made of materials such as GaAs or InP that exhibit a region of negative differential resistance. With appropriate biasing, dipole domains form and travel across the diode, allowing high frequency microwave oscillators to be built.

Light-emitting diodes (LEDs)

In a diode formed from a direct band-gap semiconductor, such as gallium arsenide, charge carriers that cross the junction emit photons when they recombine with the majority carrier on the other side. Depending on the material, wavelengths (or colors) from the infrared to the near ultraviolet may be produced. The forward potential of these diodes depends on the wavelength of the emitted photons: 2.1 V corresponds to red, 4.0 V to violet. The first LEDs were red and yellow, and higher-frequency diodes have been developed over time. All LEDs produce incoherent, narrow-spectrum light; "white" LEDs are actually combinations of three LEDs of a different color, or a blue LED with a yellow scintillator coating. LEDs can also be used as low-efficiency photodiodes in signal applications. An LED may be paired with a photodiode or phototransistor in the same package, to form an opto-isolator.

Laser diodes

When an LED-like structure is contained in a resonant cavity formed by polishing the parallel end faces, a laser can be formed. Laser diodes are commonly used in optical storage devices and for high speed optical communication.

Thermal diodes

This term is used both for conventional p–n diodes used to monitor temperature because of their varying forward voltage with temperature, and for Peltier heat pumps for thermoelectric heating and cooling. Peltier heat pumps may be made from semiconductor, though they do not have any rectifying junctions, they use the differing behaviour of charge carriers in N and P type semiconductor to move heat.

Perun's diodes

This is a special type of voltage-surge protection diode. It is characterized by the symmetrical voltage-current characteristic, similar to DIAC. It has much faster response time however, that's why it is used in demanding applications.

Photodiodes

All semiconductors are subject to optical charge carrier generation. This is typically an undesired effect, so most semiconductors are packaged in light blocking material. Photodiodes are intended to sense light(photodetector), so they are packaged in materials that allow light to pass, and are usually PIN (the kind of diode most sensitive to light). A photodiode can be used in solar cells, in photometry, or in optical communications. Multiple photodiodes may be packaged in a single device, either as a linear array or as a two-dimensional array. These arrays should not be confused with charge-coupled devices.

PIN diodes

A PIN diode has a central un-doped, or *intrinsic*, layer, forming a p-type/intrinsic/n-type structure. They are used as radio frequency switches and attenuators. They are also used as large-volume, ionizing-radiation detectors and as photodetectors. PIN diodes are also used in power electronics, as their central layer can withstand high voltages. Furthermore, the PIN structure can be found in many power semiconductor devices, such as IGBTs, power MOSFETs, and thyristors.

Schottky diodes

Schottky diodes are constructed from a metal to semiconductor contact. They have a lower forward voltage drop than p–n junction diodes. Their forward voltage drop at forward currents of about 1 mA is in the range 0.15 V to 0.45 V, which makes them useful in voltage clamping applications and prevention of transistor saturation. They can also be used as low loss rectifiers, although their reverse leakage current is in general higher than that of other diodes. Schottky diodes are majority carrier devices and so do not suffer from minority carrier storage problems that slow down many other diodes—so they have a faster reverse recovery than p–n junction diodes. They also tend to have much lower junction capacitance than p–n diodes, which provides for high switching speeds and their use in high-speed circuitry and RF devices such as switched-mode power supply, mixers, and detectors.

Super barrier diodes

Super barrier diodes are rectifier diodes that incorporate the low forward voltage drop of the Schottky diode with the surge-handling capability and low reverse leakage current of a normal p–n junction diode.

Gold-doped diodes

As a dopant, gold (or platinum) acts as recombination centers, which helps a fast recombination of minority carriers. This allows the diode to operate at signal frequencies, at the expense of a higher forward voltage drop. Gold-doped diodes are faster than other p–n diodes (but not as fast as Schottky diodes). They also have less reverse-current leakage than Schottky diodes (but not as good as other p–n diodes). A typical example is the 1N914.

Snap-off or Step recovery diodes

The term *step recovery* relates to the form of the reverse recovery characteristic of these devices. After a forward current has been passing in an SRD and the current is interrupted or reversed, the reverse conduction will cease very abruptly (as in a step waveform). SRDs can, therefore, provide very fast voltage transitions by the very sudden disappearance of the charge carriers.

Stabistors or *Forward Reference Diodes*

The term *stabistor* refers to a special type of diodes featuring extremely stable forward voltage characteristics. These devices are specially designed for low-voltage stabilization applications requiring a guaranteed voltage over a wide current range and highly stable over temperature.

Transient voltage suppression diode (TVS)

These are avalanche diodes designed specifically to protect other semiconductor devices from high-voltage transients. Their p–n junctions have a much larger cross-sectional area than those of a normal diode, allowing them to conduct large currents to ground without sustaining damage.

Varicap or varactor diodes

These are used as voltage-controlled capacitors. These are important in PLL (phase-locked loop) and FLL (frequency-locked loop) circuits, allowing tuning circuits, such as those in television receivers, to lock quickly on to the frequency. They also enabled tunable oscillators in early discrete tuning of radios, where a cheap and stable, but fixed-frequency, crystal oscillator provided the reference frequency for a voltage-controlled oscillator.

Zener diodes

These can be made to conduct in reverse bias (backward), and are correctly termed reverse breakdown diodes. This effect, called Zener breakdown, occurs at a precisely defined voltage, allowing the diode to be used as a precision voltage reference. The term Zener diode is colloquially applied to several types of breakdown diodes, but strictly speaking Zener diodes have a breakdown voltage of below 5 volts, whilst avalanche diodes are used for breakdown voltages above that value. In practical voltage reference circuits, Zener and switching diodes are connected in series and opposite directions to balance the temperature coefficient response of the diodes to near-zero. Some devices labeled as high-voltage Zener diodes are actually avalanche diodes. Two (equivalent) Zeners in series and in reverse order, in the same package, constitute a transient absorber (or Transorb, a registered trademark).

Other uses for semiconductor diodes include the sensing of temperature, and comput-ing analog logarithms.

Numbering and Coding Schemes

There are a number of common, standard and manufacturer-driven numbering and coding schemes for diodes; the two most common being the EIA/JEDEC standard and the European Pro Electron standard:

EIA/JEDEC

The standardized 1N-series numbering *EIA370* system was introduced in the US by

EIA/JEDEC (Joint Electron Device Engineering Council) about 1960. Most diodes have a 1-prefix designation (e.g., 1N4003). Among the most popular in this series were: 1N34A/1N270 (germanium signal), 1N914/1N4148 (silicon signal), 1N400x (silicon 1A power rectifier), and 1N580x (silicon 3A power rectifier).

JIS

The JIS semiconductor designation system has all semiconductor diode designations starting with "1S".

Pro Electron

The European Pro Electron coding system for active components was introduced in 1966 and comprises two letters followed by the part code. The first letter represents the semiconductor material used for the component (A = germanium and B = silicon) and the second letter represents the general function of the part (for diodes, A = low-power/ signal, B = variable capacitance, X = multiplier, Y = rectifier and Z = voltage reference); for example:

- AA-series germanium low-power/signal diodes (e.g., AA119)
- BA-series silicon low-power/signal diodes (e.g., BAT18 silicon RF switching diode)
- BY-series silicon rectifier diodes (e.g., BY127 1250V, 1A rectifier diode)
- BZ-series silicon Zener diodes (e.g., BZY88C4V7 4.7V Zener diode)

Other common numbering / coding systems (generally manufacturer-driven) include:

- GD-series germanium diodes (e.g., GD9) – this is a very old coding system
- OA-series germanium diodes (e.g., OA47) – a coding sequence developed by Mullard, a UK company

As well as these common codes, many manufacturers or organisations have their own systems too – for example:

- HP diode 1901-0044 = JEDEC 1N4148
- UK military diode CV448 = Mullard type OA81 = GEC type GEX23

Related devices

- Rectifier
- Transistor

- Thyristor or silicon controlled rectifier (SCR)
- TRIAC
- DIAC
- Varistor

In optics, an equivalent device for the diode but with laser light would be the Optical isolator, also known as an Optical Diode, that allows light to only pass in one direction. It uses a Faraday rotator as the main component.

Applications

Radio Demodulation

The first use for the diode was the demodulation of amplitude modulated (AM) radio broadcasts. The history of this discovery is treated in depth in the radio article. In summary, an AM signal consists of alternating positive and negative peaks of a radio carrier wave, whose amplitude or envelope is proportional to the original audio signal. The diode (originally a crystal diode) rectifies the AM radio frequency signal, leaving only the positive peaks of the carrier wave. The audio is then extracted from the rectified carrier wave using a simple filter and fed into an audio amplifier or transducer, which generates sound waves.

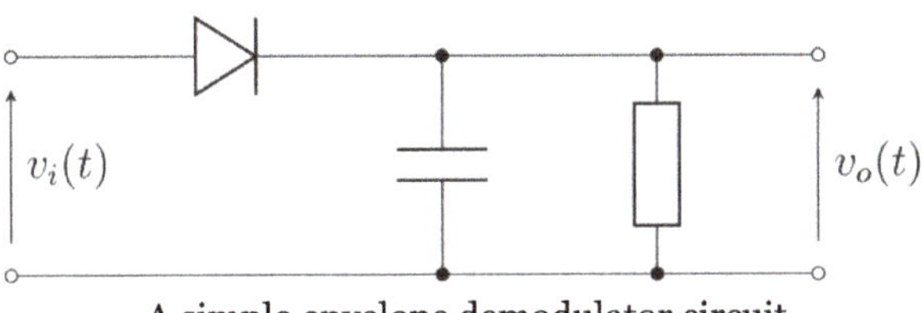

A simple envelope demodulator circuit.

Power Conversion

Rectifiers are constructed from diodes, where they are used to convert alternating current (AC) electricity into direct current (DC). Automotive alternators are a common example, where the diode, which rectifies the AC into DC, provides better performance than the commutator or earlier, dynamo. Similarly, diodes are also used in *Cockcroft–Walton voltage multipliers* to convert AC into higher DC voltages.

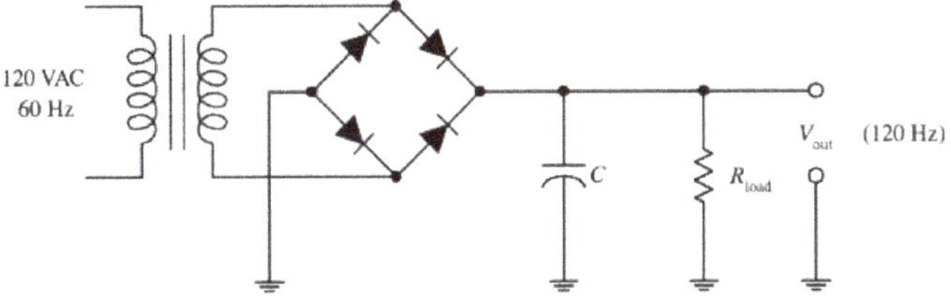

Schematic of basic AC-to-DC power supply

Over-voltage Protection

Diodes are frequently used to conduct damaging high voltages away from sensitive electronic devices. They are usually reverse-biased (non-conducting) under normal circumstances. When the voltage rises above the normal range, the diodes become forward-biased (conducting). For example, diodes are used in (stepper motor and H-bridge) motor controller and relay circuits to de-energize coils rapidly without the damaging voltage spikes that would otherwise occur. (A diode used in such an application is called a flyback diode). Many integrated circuits also incorporate diodes on the connection pins to prevent external voltages from damaging their sensitive transistors. Specialized diodes are used to protect from over-voltages at higher power.

Logic Gates

Diodes can be combined with other components to construct AND and OR logic gates. This is referred to as diode logic.

Ionizing Radiation Detectors

In addition to light, mentioned above, semiconductor diodes are sensitive to more energetic radiation. In electronics, cosmic rays and other sources of ionizing radiation cause noise pulses and single and multiple bit errors. This effect is sometimes exploited by particle detectors to detect radiation. A single particle of radiation, with thousands or millions of electron volts of energy, generates many charge carrier pairs, as its energy is deposited in the semiconductor material. If the depletion layer is large enough to catch the whole shower or to stop a heavy particle, a fairly accurate measurement of the particle's energy can be made, simply by measuring the charge conducted and without the complexity of a magnetic spectrometer, etc. These semiconductor radiation detectors need efficient and uniform charge collection and low leakage current. They are often cooled by liquid nitrogen. For longer-range (about a centimetre) particles, they need a very large depletion depth and large area. For short-range particles, they need any contact or un-depleted semiconductor on at least one surface to be very thin. The back-bias voltages are near breakdown (around a thousand volts per centimetre). Germanium and silicon are common materials. Some of these detectors sense position as well as energy. They have a finite life, especially when detecting heavy particles, because of radiation damage. Silicon and germanium are quite different in their ability to convert gamma rays to electron showers.

Semiconductor detectors for high-energy particles are used in large numbers. Because of energy loss fluctuations, accurate measurement of the energy deposited is of less use.

Temperature Measurements

A diode can be used as a temperature measuring device, since the forward voltage drop across the diode depends on temperature, as in a silicon bandgap temperature sensor.

From the Shockley ideal diode equation given above, it might *appear* that the voltage has a *positive* temperature coefficient (at a constant current), but usually the variation of the reverse saturation current term is more significant than the variation in the thermal voltage term. Most diodes therefore have a *negative* temperature coefficient, typically −2 mV/°C for silicon diodes. The temperature coefficient is approximately constant for temperatures above about 20 kelvins. Some graphs are given for 1N400x series, and CY7 cryogenic temperature sensor.

Current Steering

Diodes will prevent currents in unintended directions. To supply power to an electrical circuit during a power failure, the circuit can draw current from a battery. An uninterruptible power supply may use diodes in this way to ensure that current is only drawn from the battery when necessary. Likewise, small boats typically have two circuits each with their own battery/batteries: one used for engine starting; one used for domestics. Normally, both are charged from a single alternator, and a heavy-duty split-charge diode is used to prevent the higher-charge battery (typically the engine battery) from discharging through the lower-charge battery when the alternator is not running.

Diodes are also used in electronic musical keyboards. To reduce the amount of wiring needed in electronic musical keyboards, these instruments often use keyboard matrix circuits. The keyboard controller scans the rows and columns to determine which note the player has pressed. The problem with matrix circuits is that, when several notes are pressed at once, the current can flow backwards through the circuit and trigger "phantom keys" that cause "ghost" notes to play. To avoid triggering unwanted notes, most keyboard matrix circuits have diodes soldered with the switch under each key of the musical keyboard. The same principle is also used for the switch matrix in solid-state pinball machines.

Waveform Clipper

Diodes can be used to limit the positive or negative excursion of a signal to a prescribed voltage.

Clamper

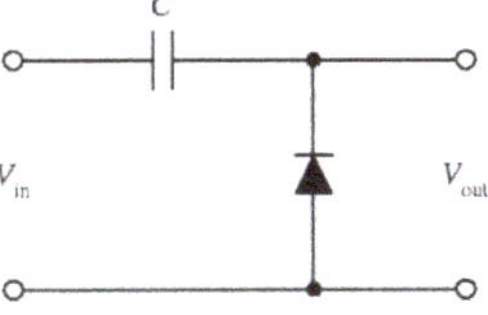

This simple diode clamp will clamp the negative peaks of the incoming waveform to the common rail voltage

A diode clamp circuit can take a periodic alternating current signal that oscillates between positive and negative values, and vertically displace it such that either the positive, or the negative peaks occur at a prescribed level. The clamper does not restrict the peak-to-peak excursion of the signal, it moves the whole signal up or down so as to place the peaks at the reference level.

Abbreviations

Diodes are usually referred to as *D* for diode on PCBs. Sometimes the abbreviation *CR* for *crystal rectifier* is used.

p–n Diode

A p–n diode is a type of semiconductor diode based upon the p–n junction. The diode conducts current in only one direction, and it is made by joining a *p*-type semiconducting layer to an *n*-type semiconducting layer. Semiconductor diodes have multiple uses including rectification of alternating current to direct current, detection of radio signals, emitting light and detecting light.

Structure

The figure shows two of the many possible structures used for *p–n* semiconductor diodes, both adapted to increase the voltage the devices can withstand in reverse bias. The top structure uses a mesa to avoid a sharp curvature of the p^+-region next to the adjoining *n*-layer. The bottom structure uses a lightly doped *p*-guard-ring at the edge of the sharp corner of the p^+-layer to spread the voltage out over a larger distance and reduce the electric field. (Superscripts like n^+ or n^- refer to heavier or lighter impurity doping levels.)

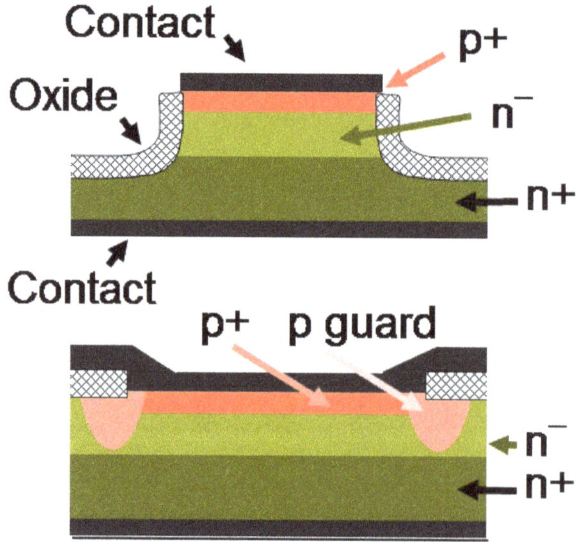

Mesa diode structure (top) and planar diode structure with guard-ring (bottom).

Electrical Behavior

The ideal diode has zero resistance for the *forward bias polarity*, and infinite resistance (conducts zero current) for the *reverse voltage polarity*; if connected in an alternating current circuit, the semiconductor diode acts as an *electrical rectifier*.

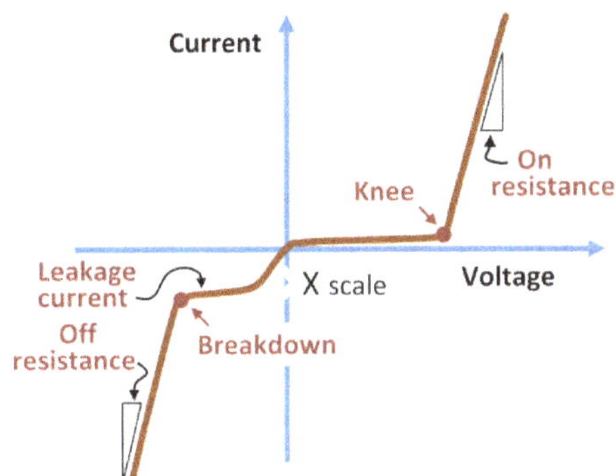

Nonideal $p–n$ diode current-voltage characteristics.

The semiconductor diode is not ideal. As shown in the figure, the diode does not conduct appreciably until a nonzero *knee voltage* (also called the *turn-on voltage* or the *cut-in voltage*) is reached. Above this voltage the slope of the current-voltage curve is not infinite (on-resistance is not zero). In the reverse direction the diode conducts a nonzero leakage current (exaggerated by a smaller scale in the figure) and at a sufficiently large reverse voltage below the *breakdown voltage* the current increases very rapidly with more negative reverse voltages.

As shown in the figure, the *on* and *off* resistances are the reciprocal slopes of the current-voltage characteristic at a selected bias point:

$$r_D = \left.\frac{\Delta v_D}{\Delta i_D}\right|_{v_D=V_{BIAS}},$$

where r_D is the resistance and Δi_D is the current change corresponding to the diode voltage change Δv_D at the bias $v_D=V_{BIAS}$.

Operation

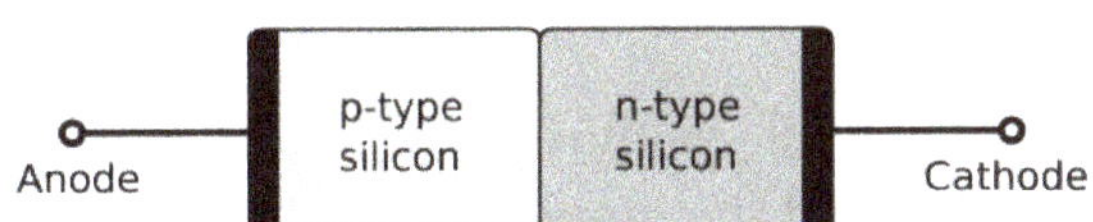

An abrupt p–n diode made by doping silicon.

Here, the operation of the abrupt $p–n$ diode is considered. By "abrupt" is meant that the p- and n-type doping exhibit a step function discontinuity at the plane where they encounter each other. The objective is to explain the various bias regimes in the figure displaying current-voltage characteristics. Operation is described using band-bending diagrams that show how the lowest conduction band energy and the highest valence band energy vary with position inside the diode under various bias conditions.

Zero Bias

The figure shows a band bending diagram for a *p*–*n* diode; that is, the band edges for the conduction band (upper line) and the valence band (lower line) are shown as a function of position on both sides of the junction between the *p*-type material (left side) and the *n*-type material (right side). When a *p*-type and an *n*-type region of the same semiconductor are brought together and the two diode contacts are short-circuited, the Fermi half-occupancy level (dashed horizontal straight line) is situated at a constant level. This level ensures that in the field-free bulk on both sides of the junction the hole and electron occupancies are correct. (So, for example, it is not necessary for an electron to leave the *n*-side and travel to the *p*-side through the short circuit to adjust the occupancies.)

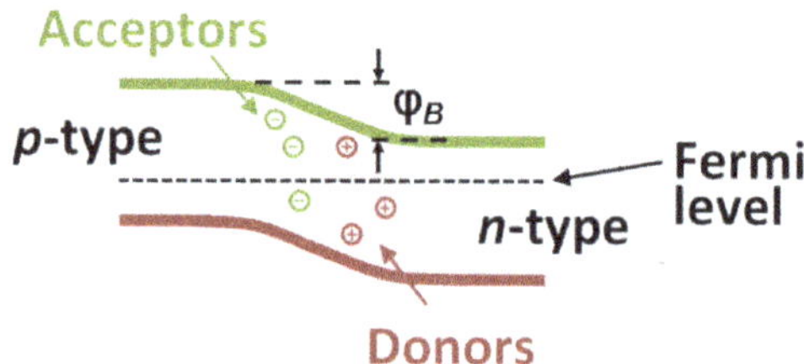

Band-bending diagram for *p*–*n* diode at zero applied voltage. The depletion region is shaded.

However, a flat Fermi level requires the bands on the *p*-type side to move higher than the corresponding bands on the *n*-type side, forming a step or barrier in the band edges, labeled φ_B. This step forces the electron density on the *p*-side to be a Boltzmann factor $\exp(-\varphi_B/V_{th})$ smaller than on the *n*-side, corresponding to the lower electron density in *p*-region. The symbol V_{th} denotes the *thermal voltage*, defined as $V_{th} = k_B T/q$. At T = 290 kelvins (room temperature), the thermal voltage is approximately 25 mV. Similarly, hole density on the *n*-side is a Boltzmann factor smaller than on the *p*-side. This reciprocal reduction in minority carrier density across the junction forces the *pn*-product of carrier densities to be

$$pn = p_B n_B e^{-\varphi_B/V_{th}}$$

at any position within the diode at equilibrium. Where p_B and n_B are the bulk majority carrier densities on the *p*-side and the *n*-side, respectively.

As a result of this step in band edges, a *depletion region* near the junction becomes depleted of both holes and electrons, forming an insulating region with almost no *mobile* charges. There are, however, *fixed, immobile* charges due to dopant ions. The near absence of mobile charge in the depletion layer means that the mobile charges present are insufficient to balance the immobile charge contributed by the dopant ions: a negative charge on the *p*-type side due to acceptor dopant and as a positive charge on the *n*-type side due to donor dopant. Because of this charge there is an electric field in this region, as determined by Poisson's equation. The width of the depletion region adjusts so the negative acceptor charge on the p-side exactly balances the positive donor charge on the n-side, so there is no electric field outside the depletion region on either side.

In this band configuration no voltage is applied and no current flows through the diode. To force current through the diode a *forward bias* must be applied, as described next.

Forward Bias

In forward bias, positive terminal of the battery is connected to the p- type material and negative terminal is connected to the n- type material so that holes are injected into the p-type material and electrons into the n-type material. The electrons in the n-type material are called *majority* carriers on that side, but electrons that make it to the p-type side are called *minority* carriers. The same descriptors apply to holes: they are majority carriers on the p-type side, and minority carriers on the n-type side.

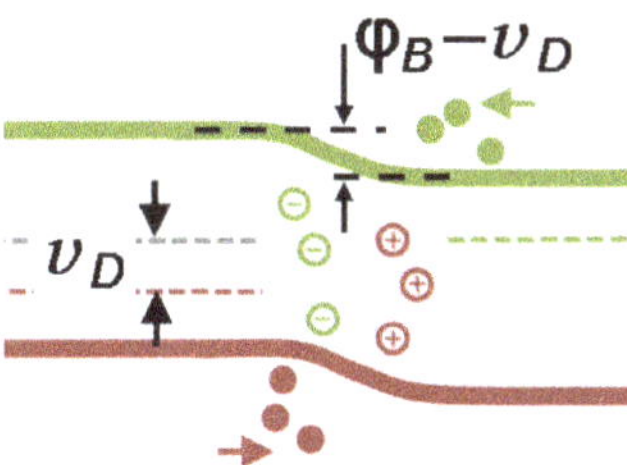

Band-bending diagram for p–n diode in forward bias. Diffusion drives carriers across the junction.

A forward bias separates the two bulk half-occupancy levels by the amount of the applied voltage, which lowers the separation of the p-type bulk band edges to be closer in energy to those of the n-type. As shown in the diagram, the step in band edges is reduced by the applied voltage to $\varphi_B - v_D$. (The band bending diagram is made in units of volts, so no electron charge appears to convert v_D to energy.)

Under forward bias, a *diffusion current* flows (that is a current driven by a concentration gradient) of holes from the p-side into the n-side, and of electrons in the opposite direction from the n-side to the p-side. The gradient driving this transfer is set up as follows: in the bulk distant from the interface, minority carriers have a very low concentration compared to majority carriers, for example, electron density on the p-side (where they are minority carriers) is a factor $\exp(-\varphi_B/V_{th})$ lower than on the n-side (where they are majority carriers). On the other hand, near the interface, application of voltage v_D reduces the step in band edges and increases minority carrier densities by a Boltzmann factor $\exp(v_D/V_{th})$ above the bulk values. Within the junction, the pn-product is increased above the equilibrium value to:

$$pn = \left(p_B n_B \, e^{-\varphi_B/V_{th}} \right) e^{v_D/V_{th}} \, .$$

The gradient driving the diffusion is then the difference between the large excess minority carrier densities at the barrier and the low densities in the bulk, and that gradient drives diffusion of minority carriers from the interface into the bulk. The injected minority carriers are reduced in number as they travel into the bulk by *recombination* mechanisms that drive the excess concentrations toward the bulk values.

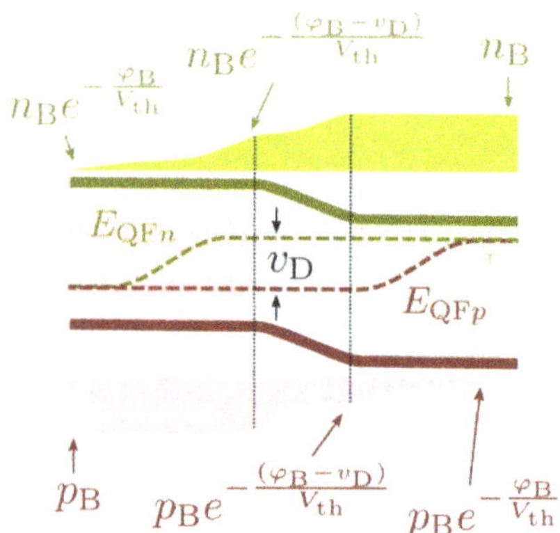

Quasi-Fermi levels and carrier densities in forward biased p–n- diode. The figure assumes recombination is confined to the regions where majority carrier concentration is near the bulk values, which is not accurate when recombination-generation centers in the field region play a role.

Recombination can occur by direct encounter with a majority carrier, annihilating both carriers, or through a *recombination-generation* center, a defect that alternately traps holes and electrons, assisting recombination. The minority carriers have a limited *lifetime*, and this lifetime in turn limits how far they can diffuse from the majority carrier side into the minority carrier side, the so-called *diffusion length*. In the LED recombination of electrons and holes is accompanied by emission of light of a wavelength related to the energy gap between valence and conduction bands, so the diode converts a portion of the forward current into light.

Under forward bias, the half-occupancy lines for holes and electrons cannot remain flat throughout the device as they are when in equilibrium, but become *quasi-Fermi levels* that vary with position. As shown in the figure, the electron quasi-Fermi level shifts with position, from the half-occupancy equilibrium Fermi level in the n-bulk, to the half-occupancy equilibrium level for holes deep in the p-bulk. The hole quasi-Fermi level does the reverse. The two quasi-Fermi levels do not coincide except deep in the bulk materials.

The figure shows the majority carrier densities drop from the majority carrier density levels n_B, p_B in their respective bulk materials, to a level a factor $\exp(-(\varphi_B - v_D)/V_{th})$ smaller at the top of the barrier, which is reduced from the equilibrium value φ_B by the amount of the forward diode bias v_D. Because this barrier is located in the oppositely doped material, the injected carriers at the barrier position are now minority carriers. As recombination takes hold, the minority carrier densities drop with depth to their equilibrium values for bulk minority carriers, a factor $\exp(-\varphi_B/V_{th})$ smaller than their bulk densities n_B, p_B as majority carriers before injection. At this point the quasi-Fermi levels rejoin the bulk Fermi level positions.

The reduced step in band edges also means that under forward bias the depletion region narrows as holes are pushed into it from the *p*-side and electrons from the *n*-side.

In the simple *p*–*n* diode the forward current increases exponentially with forward bias voltage due to the exponential increase in carrier densities, so there is always some current at even very small values of applied voltage. However, if one is interested in some particular current level, it will require a "knee" voltage before that current level is reached. For example, a very common choice in texts about circuits using silicon diodes is V_{Knee} = 0.7 V. Above the knee, the current continues to increase exponentially. Some special diodes, such as some varactors, are designed deliberately to maintain a low current level up to some knee voltage in the forward direction.

Reverse Bias

In reverse bias the occupancy level for holes again tends to stay at the level of the bulk *p*-type semiconductor while the occupancy level for electrons follows that for the bulk *n*-type. In this case, the *p*-type bulk band edges are raised relative to the *n*-type bulk by the reverse bias v_R, so the two bulk occupancy levels are separated again by an energy determined by the applied voltage. As shown in the diagram, this behavior means the step in band edges is increased to φ_B+v_R, and the depletion region widens as holes are pulled away from it on the *p*-side and electrons on the *n*-side.

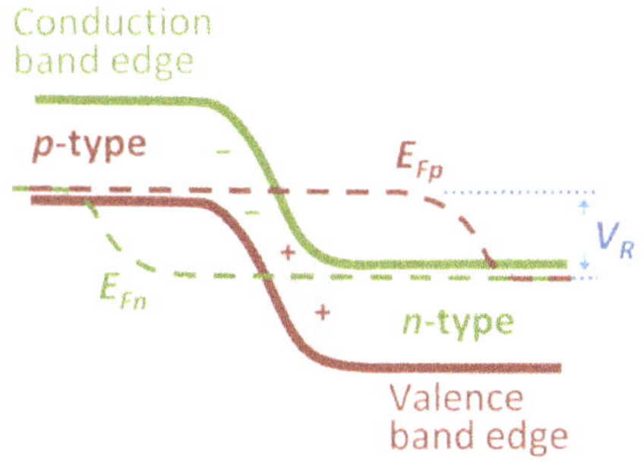

Quasi-Fermi levels in reverse-biased *p*–*n* diode.

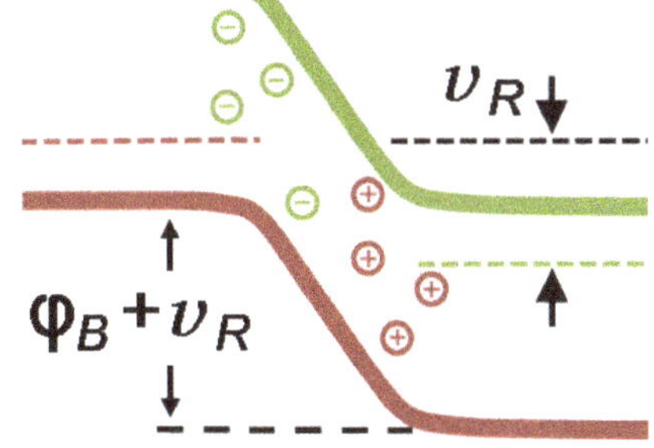

Band-bending for *p*–*n* diode in reverse bias

When the reverse bias is applied, the electric field in the depletion region is increased, pulling the electrons and holes further apart than in the zero bias case. Thus, any current that flows is due to the very weak process of carrier generation inside the depletion region due to *generation-recombination defects* in this region. That very small current is the source of the leakage current under reverse bias. In the photodiode, reverse current is introduced using creation of holes and electrons in the depletion region by incident light, thus converting a portion of the incident light into an electric current.

When the reverse bias becomes very large, reaching the breakdown voltage, the generation process in the depletion region accelerates leading to an *avalanche* condition which can cause runaway and destroy the diode.

Diode Law

The DC current-voltage behavior of the ideal *p–n* diode is governed by the Shockley diode equation:

$$i_{\mathrm{D}} = I_{\mathrm{R}}\left(e^{v_{\mathrm{D}}/V_{\mathrm{th}}} - 1\right),$$

where v_{D} is the DC voltage across the diode and I_{R} is the *reverse saturation current*, the current that flows when the diode is reverse biased (that is, v_{D} is large and negative). The quantity V_{th} is the *thermal voltage* defined as $V_{\mathrm{th}} = k_{\mathrm{B}}T/q$. This is approximately equal to 25 mV at $T = 290$ kelvins.

This equation does not model the non-ideal behavior such as excess reverse leakage or breakdown phenomena. In many practical diodes this equation must be modified to read

$$i_{D} = I_{R}\left(e^{v_{\mathrm{D}}/nV_{\mathrm{th}}} - 1\right),$$

where n is an *ideality factor* introduced to model a slower rate of increase than predicted by the ideal diode law. Using this equation, the diode *on*-resistance is

$$r_{\mathrm{D}} = \frac{1}{di_{\mathrm{D}}/dv_{\mathrm{D}}} \approx \frac{nV_{\mathrm{th}}}{i_{\mathrm{D}}},$$

exhibiting a lower resistance the higher the current.

Capacitance

The depletion layer between the *n*- and *p*-sides of a *p–n*-diode serves as an insulating region that separates the two diode contacts. Thus, the diode in reverse bias exhibits a *depletion-layer capacitance*, sometimes more vaguely called a *junction capacitance*, analogous to a parallel plate capacitor with a dielectric spacer between the contacts. In reverse bias the width of the depletion layer is widened with increasing reverse bias v_R, and the capacitance is accordingly decreased. Thus, the junction serves as a voltage-controllable capacitor. In a simplified one-dimensional model, the junction capacitance is:

$$C_J = \kappa\varepsilon_0 \frac{A}{w(v_R)},$$

with A the device area, κ the relative semiconductor dielectric permittivity, ε_o the electric constant, and w the depletion width (thickness of the region where mobile carrier density is negligible).

In forward bias, besides the above depletion-layer capacitance, minority carrier charge injection and diffusion occurs. A *diffusion capacitance* exists expressing the change in minority carrier charge that occurs with a change in forward bias. In terms of the stored minority carrier charge, the diode current i_D is:

$$i_D = \frac{Q_D}{\tau_T},$$

where Q_D is the charge associated with diffusion of minority carriers, and τ_T is the *transit time*, the time taken for the minority charge to transit the injection region. Typical values for transit time are 0.1–100 ns. On this basis, the diffusion capacitance is calculated to be:

$$C_D = \frac{dQ_D}{dv_D} = \tau_T \frac{di_D}{dv_D} = \frac{i_D \tau_T}{V_{th}}.$$

Generally speaking, for usual current levels in forward bias, this capacitance far exceeds the depletion-layer capacitance.

Transient Response

The diode is a highly non-linear device, but for small-signal variations its response can be analyzed using a *small-signal circuit* based upon the DC bias about which the signal is imagined to vary. The equivalent circuit is shown at the right for a diode driven by a Norton source. Using Kirchhoff's current law at the output node:

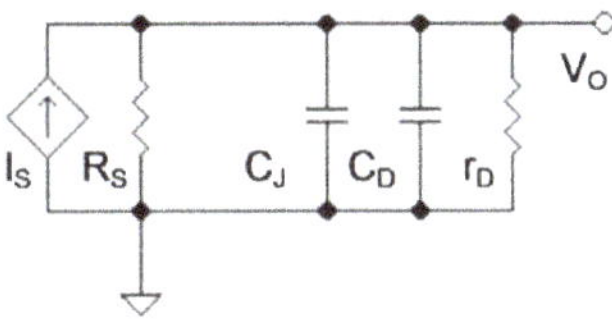

Small-signal circuit for *p–n* diode driven by a current signal represented as a Norton source.

$$I_S = \left(j\omega(C_J + C_D) + \frac{1}{r_D} + \frac{1}{R_S} \right) V_O,$$

with C_D the diode diffusion capacitance, C_J the diode junction capacitance (the depletion layer capacitance) and r_D the diode resistance, all at the selected quiescent bias point or Q-point. The output voltage provided by this circuit is then:

$$\frac{V_O}{I_S} = \frac{(R_S \,||\, r_D)}{1 + j\omega(C_D + C_J)(R_S \,||\, r_D)},$$

with $(R_S||r_D)$ the parallel combination of R_S and r_D. This *transresistance amplifier* exhibits a *corner frequency*, denoted f_C:

$$f_C = \frac{1}{2\pi(C_D + C_J)(R_S \parallel r_D)},$$

and for frequencies $f >> f_C$ the gain rolls off with frequency as the capacitors short-circuit the resistor r_D. Assuming, as is the case when the diode is turned on, that $C_D >> C_J$ and $R_S >> r_D$, the expressions found above for the diode resistance and capacitance provide:

$$f_C = \frac{1}{2\pi n \tau_T},$$

which relates the corner frequency to the diode transit time τ_T.

For diodes operated in reverse bias, C_D is zero and the term *corner frequency* often is replaced by *cutoff frequency*. In any event, in reverse bias the diode resistance becomes quite large, although not infinite as the ideal diode law suggests, and the assumption that it is less than the Norton resistance of the driver may not be accurate. The junction capacitance is small and depends upon the reverse bias v_R. The cutoff frequency is then:

$$f_C = \frac{1}{2\pi C_J(R_S \parallel r_D)},$$

and varies with reverse bias because the width $w(v_R)$ of the insulating region depleted of mobile carriers increases with increasing diode reverse bias, reducing the capacitance.

PIN Diode

A PIN diode is a diode with a wide, undoped intrinsic semiconductor region between a p-type semiconductor and an n-type semiconductor region. The p-type and n-type regions are typically heavily doped because they are used for ohmic contacts.

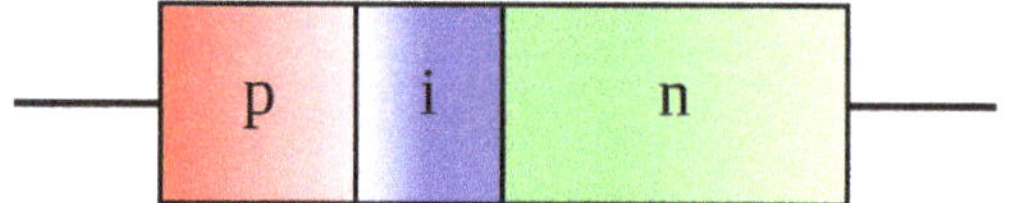

Layers of a PIN diode

The wide intrinsic region is in contrast to an ordinary p–n diode. The wide intrinsic region makes the PIN diode an inferior rectifier (one typical function of a diode), but it makes it suitable for attenuators, fast switches, photodetectors, and high voltage power electronics applications.

Operation

A PIN diode operates under what is known as high-level injection. In other words, the intrinsic "i" region is flooded with charge carriers from the "p" and "n" regions. Its function can be likened to filling up a water bucket with a hole on the side. Once the water reaches the hole's level it will begin to pour out. Similarly, the diode will conduct current once the flooded electrons and holes reach an equilibrium point, where the number of electrons is equal to the number of holes in the intrinsic region. When the diode is forward biased, the injected carrier concentration is typically several orders of magnitude higher than the intrinsic carrier concentration. Due to this high level injection, which in turn is due to the depletion process, the electric field extends deeply (almost the entire length) into the region. This electric field helps in speeding up of the transport of charge carriers from the P to the N region, which results in faster operation of the diode, making it a suitable device for high frequency operations.

Characteristics

A PIN diode obeys the standard diode equation for low frequency signals. At higher frequencies, the diode looks like an almost perfect (very linear, even for large signals) resistor. There is a lot of stored charge in the intrinsic region. At low frequencies, the charge can be removed and the diode turns off. At higher frequencies, there is not enough time to remove the charge, so the diode never turns off. The PIN diode has a poor reverse recovery time.

The high-frequency resistance is inversely proportional to the DC bias current through the diode. A PIN diode, suitably biased, therefore acts as a variable resistor. This high-frequency resistance may vary over a wide range (from 0.1 Ω to 10 kΩ in some cases; the useful range is smaller, though).

The wide intrinsic region also means the diode will have a low capacitance when reverse-biased.

In a PIN diode, the depletion region exists almost completely within the intrinsic region. This depletion region is much larger than in a PN diode, and almost constant-size, independent of the reverse bias applied to the diode. This increases the volume where electron-hole pairs can be generated by an incident photon. Some photodetector devices, such as PIN photodiodes and phototransistors (in which the base-collector junction is a PIN diode), use a PIN junction in their construction.

The diode design has some design trade-offs. Increasing the dimensions of the intrinsic region (and its stored charge) allows the diode to look like a resistor at lower frequencies. It adversely affects the time needed to turn off the diode and its shunt capacitance. It is therefore necessary to select a device with the appropriate properties for a particular use.

Applications

PIN diodes are useful as RF switches, attenuators, photodetectors, and phase shifters.

RF and Microwave Switches

Under zero- or reverse-bias (the "off" state), a PIN diode has a low capacitance. The low capacitance will not pass much of an RF signal. Under a forward bias of 1 mA (the "on" state), a typical PIN diode will have an RF resistance of about 1 ohm, making it a good RF conductor. Consequently, the PIN diode makes a good RF switch.

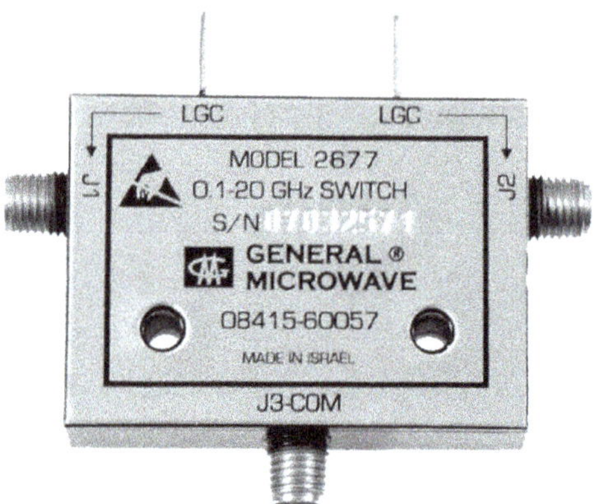

A PIN Diode RF Microwave Switch.

Although RF relays can be used as switches, they switch very slowly (on the order of 10 milliseconds). A PIN diode switch can switch much more quickly (e.g., 1 microsecond).

For example, the capacitance of an "off"-state discrete PIN diode might be 1 pF. At 320 MHz, the capacitive reactance of 1 pF is 497 ohms:

$$Z_{diode} = \frac{1}{2\pi fC} = \frac{1}{2 \cdot \pi \cdot 320\times10^{6} \cdot 1\times10^{-12}} = 497\ \Omega$$

As a series element in a 50 ohm system, the off-state attenuation in dB is:

$$A = 20\log_{10}\left(\frac{Z_{load}}{Z_{source} + Z_{diode} + Z_{load}}\right) = 20\log_{10}\left(\frac{50}{50+497+50}\right) = 21.5\ dB$$

This attenuation may not be adequate. In applications where higher isolation is needed, both shunt and series elements may be used, with the shunt diodes biased in complementary fashion to the series elements. Adding shunt elements effectively reduces the source and load impedances, reducing the impedance ratio and increasing the off-state attenuation. However, in addition to the added complexity, the on-state attenuation is increased due to the series resistance of the on-state blocking element and the capacitance of the off-state shunt elements.

PIN diode switches are used not only for signal selection, but also component selection. For example, some low phase noise oscillators use them to range-switch inductors.

RF and Microwave Variable Attenuators

By changing the bias current through a PIN diode, it is possible to quickly change the RF resistance.

An RF Microwave PIN diode Attenuator.

At high frequencies, the PIN diode appears as a resistor whose resistance is an inverse function of its forward current. Consequently, PIN diode can be used in some variable attenuator designs as amplitude modulators or output leveling circuits.

PIN diodes might be used, for example, as the bridge and shunt resistors in a bridged-T attenuator. Another common approach is to use PIN diodes as terminations connected to the 0 degree and -90 degree ports of a quadrature hybrid. The signal to be attenuated is applied to the input port, and the attenuated result is taken from the isolation port. The advantages of this approach over the bridged-T and pi approaches are (1) complementary PIN diode bias drives are not needed—the same bias is applied to both diodes—and (2) the loss in the attenuator equals the return loss of the terminations, which can be varied over a very wide range.

Limiters

PIN diodes are sometimes used as input protection devices for high frequency test probes. If the input signal is within range, the PIN diode has little impact as a small capacitance. If the signal is large, then the PIN diode starts to conduct and becomes a resistor that shunts most of the signal to ground.

Photodetector and Photovoltaic Cell

The PIN photodiode was invented by Jun-ichi Nishizawa and his colleagues in 1950.

PIN photodiodes are used in fibre optic network cards and switches. As a photodetector, the PIN diode is reverse-biased. Under reverse bias, the diode ordinarily does not conduct (save a small dark current or I_s leakage). When a photon of sufficient energy enters the depletion region of the diode, it creates an electron, hole pair. The reverse bias field sweeps the carriers out of the region creating a current. Some detectors can use avalanche multiplication.

The same mechanism applies to the PIN structure, or p-i-n junction, of a solar cell. In this case, the advantage of using a PIN structure over conventional semiconductor p–n junction is the better long wavelength response of the former. In case of long wavelength irradiation, photons penetrate deep into the cell. But only those electron-hole pairs generated in and near the depletion region contribute to current generation. The depletion region of a PIN structure extends across the intrinsic region, deep into the device. This wider depletion width enables electron-hole pair generation deep within the device. This increases the quantum efficiency of the cell.

Commercially available PIN photodiodes have quantum efficiencies above 80-90% in the telecom wavelength range (~1500 nm), and are typically made of germanium or InGaAs. They feature fast response times (higher than their p-n counterparts), running into several tens of gigahertz, making them ideal for high speed optical telecommunication applications. Similarly, silicon p-i-n photodiodes have even higher quantum efficiencies, but can only detect wavelengths below the bandgap of silicon, i.e. ~1100 nm.

Typically, amorphous silicon thin-film cells use PIN structures. On the other hand, CdTe cells use NIP structure, a variation of the PIN structure. In a NIP structure, an intrinsic CdTe layer is sandwiched by n-doped CdS and p-doped ZnTe. The photons are incident on the n-doped layer unlike a PIN diode.

A PIN photodiode can also detect X-ray and gamma ray photons.

Example Diodes

SFH203 and BPW43 are cheap general purpose PIN diodes in 5 mm clear plastic cases with bandwidths over 100 MHz. RONJA telecommunication systems are an example application.

Depletion Region

In semiconductor physics, the depletion region, also called depletion layer, depletion zone, junction region, space charge region or space charge layer, is an insulating region within a conductive, doped semiconductor material where the mobile charge carriers have been diffused away, or have been forced away by an electric field. The only elements left in the depletion region are ionized donor or acceptor impurities.

The depletion region is so named because it is formed from a conducting region by removal of all free charge carriers, leaving none to carry a current. Understanding the depletion region is key to explaining modern semiconductor electronics: diodes, bipolar junction transistors, field-effect transistors, and variable capacitance diodes all rely on depletion region phenomena.

The following discussion is limited to the p–n junction and the MOS capacitor, but depletion regions arise in all the devices mentioned above.

Formation of Depletion Region in a p–n Junction

A depletion region forms instantaneously across a p–n junction. It is most easily described when the junction is in thermal equilibrium or in a steady state: in both of these cases the properties of the system do not vary in time; they have been called dynamic equilibrium.

Electrons and holes diffuse into regions with lower concentrations of electrons and holes, much as ink diffuses into water until it is uniformly distributed. By definition, N-type semiconductor has an excess of free electrons compared to the P-type region, and P-type has an excess of holes compared to the N-type region. Therefore, when N-doped and P-doped pieces of semiconductor are placed together to form a junction, electrons migrate into the P-side and holes migrate into the N-side. Departure of an electron from the N-side to the P-side leaves a positive donor ion behind on the N-side, and likewise the hole leaves a negative acceptor ion on the P-side.

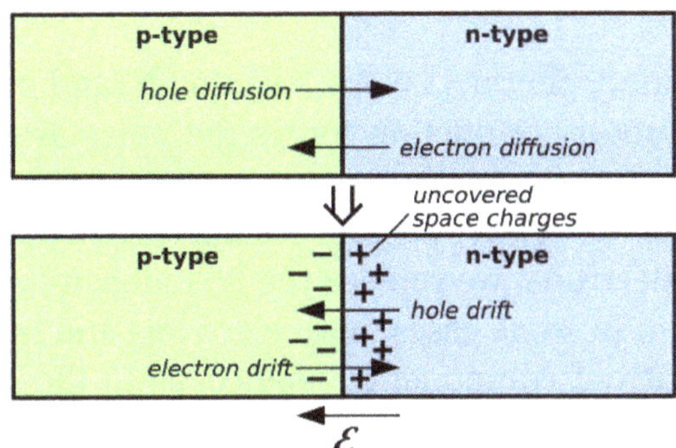

Top: p–n junction before diffusion; Bottom: After equilibrium is reached

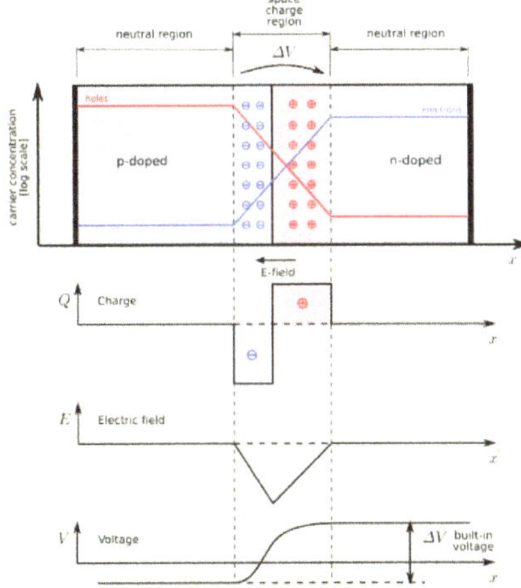

Top: hole and electron concentrations through the junction; Second from top: charge densities; Third: electric field; Bottom: electric potential

Following transfer, the diffused electrons come into contact with holes on the P-side and are eliminated by recombination. Likewise for the diffused holes on the N-side. The net result is the diffused electrons and holes are gone, leaving behind the charged ions adjacent to the interface in a region with no mobile carriers (That's why it is called the depletion region; carriers are being depleted). The uncompensated ions are positive on the N side and negative on the P side. This creates an electric field that provides a force opposing the continued exchange of charge carriers. When the electric field is sufficient to arrest further transfer of holes and electrons, the depletion region has reached its equilibrium dimensions. Integrating the electric field across the depletion region determines what is called the built-in voltage (also called the junction voltage or barrier voltage or contact potential).

> Mathematically speaking, charge transfer in semiconductor devices is due both to conduction driven by the electric field (drift) and by diffusion. For a P-type region, where holes conduct with electrical conductivity σ and diffuse with diffusion constant D, the net current density is given by
>
> $$j = \sigma E - D \nabla qp$$
>
> with q the elementary charge (1.6×10^{-19} coulomb) and p the hole density (number per unit volume). Conduction forces the holes along the direction of the electric field. Diffusion moves the carriers in the direction of decreasing concentration, so for holes a negative current results for a positive density gradient. (If the carriers are electrons, we replace the hole density p by the negative of the electron density n; in some cases, both electrons and holes must be included.) When the two current components balance, as in the p–n-junction depletion region at dynamic equilibrium, the current is zero due to the Einstein relation, which relates D to σ.

Forward Bias

Forward bias (P positive with respect to N) narrows the depletion region and lowers the barrier to carrier injection (shown in the figure to the right). In more detail, majority carriers get some energy from the bias field, enabling them to go into the region and neutralize opposite charges. The more bias the more neutralization (or screening of ions in the region) occurs. The carriers can be recombined to the ions but thermal energy immediately makes recombined carriers transition back as Fermi energy is in proximity. When bias is strong enough that the depletion region becomes very thin, the diffusion component of the current greatly increases and the drift component increases. In this case the net current is rightward in the figure of the p–n junction. The carrier density is large (it varies exponentially with the applied bias voltage), making the junction conductive and allowing a large forward current. The mathematical description of the current is provided by the Shockley diode equation. The low current conducted under reverse bias and the large current under forward bias is an example of rectification.

Reverse Bias

Under reverse bias (P negative with respect to N), the potential drop (i.e., voltage) across the depletion region increases. In more detail, majority carriers are pushed away from the junction, leaving behind more charged ions. Thus the depletion region is widened and its field becomes stronger, which increases the drift component of current and decreases the diffusion component. In this case, the net current is leftward in the figure of the p–n junction. The carrier density (mostly, minority carriers) is small and only a very small *reverse saturation current* flows.

Formation of Depletion Region in an MOS Capacitor

Another example of a depletion region occurs in the MOS capacitor. It is shown in the figure, for a P-type substrate. Suppose that the semiconductor initially is charge neutral, with the charge due to holes exactly balanced by the negative charge due to acceptor doping impurities. If a positive voltage now is applied to the gate, which is done by introducing positive charge Q to the gate, then some positively charged holes in the semiconductor nearest the gate are repelled by the positive charge on the gate, and exit the device through the bottom contact. They leave behind a *depleted* region that is insulating because no mobile holes remain; only the immobile, negatively charged acceptor impurities. The greater the positive charge placed on the gate, the more positive the applied gate voltage, and the more holes that leave the semiconductor surface, enlarging the depletion region. (In this device there is a limit to how wide the depletion width may become. It is set by the onset of an inversion layer of carriers in a thin layer, or channel, near the surface. The above discussion applies for positive voltages low enough that an inversion layer does not form.)

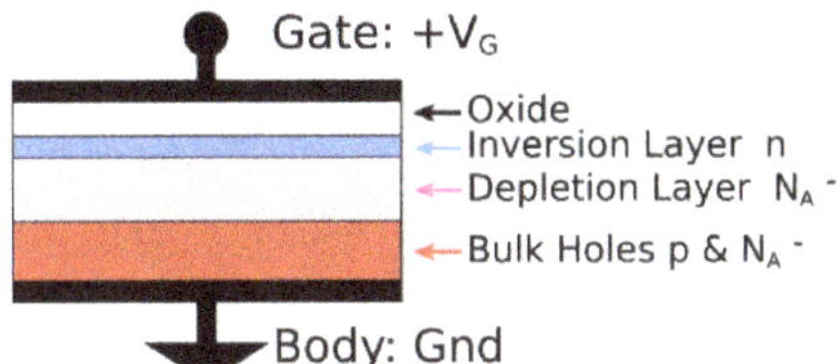

Metal–oxide–semiconductor structure on P-type silicon

If the gate material is polysilicon of opposite type to the bulk semiconductor, then a spontaneous depletion region forms if the gate is electrically shorted to the substrate, in much the same manner as described for the p–n junction above.

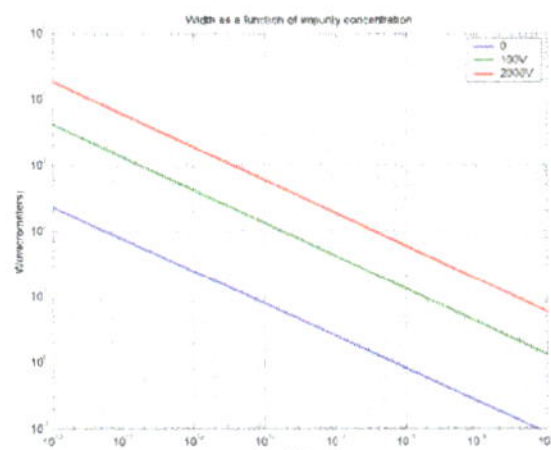

The total width of the depletion region is a function of applied reverse-bias and impurity concentration

The principle of charge neutrality says the sum of positive charges must equal the sum of negative charges:

$$n + N_A = p + N_D ,$$

where n and p are the number of free electrons and holes, and N_D and N_A are the number of ionized donors and acceptors, respectively. If we assume full ionization and that $n, p << N_D, N_A$, then:

$$qN_A w_P \approx qN_D w_N ..$$

This condition ensures that the net negative acceptor charge exactly balances the net positive donor charge. The total depletion width in this case is the sum . A full derivation for the depletion width is presented in reference. This derivation is based on solving the Poisson equation in one dimension – the dimension normal to the metallurgical junction. The electric field is zero outside of the depletion width and therefore Gauss's law implies that the charge density in each region balance – as shown by the first equation in this sub-section. Treating each region separately and substituting the charge density for each region into the Poisson equation eventually leads to a result for the depletion width. This result for the depletion width is:

$$W \approx \left[\frac{2\epsilon_r \epsilon_0}{q} \left(\frac{N_A + N_D}{N_A N_D} \right) \left(V_{bi} - V \right) \right]^{\frac{1}{2}}$$

where ϵ_r is the relative dielectric permittivity of the semiconductor, V_{bi} is the built-in voltage, and V is the applied bias. The depletion region is not symmetrically split between the n and p regions - it will tend towards the lightly doped side. A more complete analysis would take into account that there are still *some* carriers near the edges of the depletion region. This leads to an additional -2kT/q term in the last set of parentheses above.

Depletion Width in MOS Capacitor

As in p–n junctions, the governing principle here is charge neutrality. Let us assume a P-type substrate. If positive charge Q is placed on the gate, then holes are depleted to a depth w exposing sufficient negative acceptors to exactly balance the gate charge. Supposing the dopant density to be N_A acceptors per unit volume, then charge neutrality requires the depletion width w to satisfy the relationship:

$$Q = qN_A w$$

If the depletion width becomes wide enough, then electrons appear in a very thin layer at the semiconductor-oxide interface, called an inversion layer because they are oppositely charged to the holes that prevail in a P-type material. When an inversion layer forms the depletion width ceases to expand with increase in gate charge Q. In this case neutrality is achieved by attracting more electrons into the inversion layer. In the MOSFET this inversion layer is referred to as the channel.

Electric Field in Depletion Layer and Band Bending

Associated with the depletion layer is an effect known as band bending. This occurs because the electric field in the depletion layer varies linearly in space from its (maximum) value E_m at the gate to zero at the edge of the depletion width:

$$E_m = Q / A\epsilon_0 = qN_A w / A\epsilon_0,$$

where A is the gate area, $\epsilon_0 = 8.854\times10^{-12}$ F/m, F is the farad and m is the meter. This linearly-varying electric field leads to an electrical potential that varies quadratically in space. The energy levels, or energy bands, *bend* in response to this potential.

Photoelectric Effect

The photoelectric effect or *photoemission* (given by *Albert Einstein)* is the production of electrons or other free carriers when light is shone onto a material. Electrons emitted in this manner can be called *photoelectrons*. The phenomenon is commonly studied in electronic physics, as well as in fields of chemistry, such as quantum chemistry or electrochemistry.

According to classical electromagnetic theory, this effect can be attributed to the transfer of energy from the light to an electron. From this perspective, an alteration in the intensity of light would induce changes in the rate of emission of electrons from the metal. Furthermore, according to this theory, a sufficiently dim light would be expected to show a time lag between the initial shining of its light and the subsequent emission of an electron. However, the experimental results did not correlate with either of the two predictions made by classical theory.

Instead, electrons are dislodged only by the impingement of photons when those photons reach or exceed a threshold frequency (energy). Below that threshold, no electrons are emitted from the metal regardless of the light intensity or the length of time of exposure to the light. To make sense of the fact that light can eject electrons even if its intensity is low, Albert Einstein proposed that a beam of light is not a wave propagating through space, but rather a collection of discrete wave packets (photons), each with energy hf. This shed light on Max Planck's previous discovery of the Planck relation ($E = hf$) linking energy (E) and frequency (f) as arising from quantization of energy. The factor h is known as the Planck constant.

In 1887, Heinrich Hertz discovered that electrodes illuminated with ultraviolet light create electric sparks more easily. In 1905 Albert Einstein published a paper that explained experimental data from the photoelectric effect as the result of light energy being carried in discrete quantized packets. This discovery led to the quantum revolution. In 1914, Robert Millikan's experiment confirmed Einstein's law on photoelectric effect. Einstein was awarded the Nobel Prize in 1921 for "his discovery of the law of the photoelectric effect", and Millikan was awarded the Nobel Prize in 1923 for "his work on the elementary charge of electricity and on the photoelectric effect".

The photoelectric effect requires photons with energies approaching zero (in the case of negative electron affinity) to over 1 MeV for core electrons in elements with a high atomic number. Emission of conduction electrons from typical metals usually requires a few electron-volts, corresponding to short-wavelength visible or ultraviolet light. Study of the photoelectric effect led to important steps in understanding the quantum nature of light and electrons and influenced the formation of the concept of wave–particle duality. Other phenomena where light affects the movement of electric charges include the photoconductive effect (also known as photoconductivity or photoresistivity), the photovoltaic effect, and the photoelectrochemical effect.

Photoemission can occur from any material, but it is most easily observable from metals or other conductors because the process produces a charge imbalance, and if this charge imbalance is not neutralized by current flow (enabled by conductivity), the potential barrier to emission increases until the emission current ceases. It is also usual to have the emitting surface in a vacuum, since gases impede the flow of photoelectrons and make them difficult to observe. Additionally, the energy barrier to photoemission is usually increased by thin oxide layers on metal surfaces if the metal has been exposed to oxygen, so most practical experiments and devices based on the photoelectric effect use clean metal surfaces in a vacuum.

When the photoelectron is emitted into a solid rather than into a vacuum, the term *internal photoemission* is often used, and emission into a vacuum distinguished as *external photoemission*.

Emission Mechanism

The photons of a light beam have a characteristic energy proportional to the frequency of the light. In the photoemission process, if an electron within some material absorbs the energy of one photon and acquires more energy than the work function (the electron binding energy) of the material, it is ejected. If the photon energy is too low, the electron is unable to escape the material. Since an increase in the intensity of low-frequency light will only increase the number of low-energy photons sent over a given interval of time, this change in intensity will not create any single photon with enough energy to dislodge an electron. Thus, the energy of the emitted electrons does not depend on the intensity of the incoming light, but only on the energy (equivalently frequency) of the individual photons. It is an interaction between the incident photon and the outermost electrons.

Electrons can absorb energy from photons when irradiated, but they usually follow an "all or nothing" principle. All of the energy from one photon must be absorbed and used to liberate one electron from atomic binding, or else the energy is re-emitted. If the photon energy is absorbed, some of the energy liberates the electron from the atom, and the rest contributes to the electron's kinetic energy as a free particle.

Experimental Observations of Photoelectric Emission

The theory of the photoelectric effect must explain the experimental observations of the emission of electrons from an illuminated metal surface.

For a given metal, there exists a certain minimum frequency of incident radiation below which no photoelectrons are emitted. This frequency is called the *threshold frequency*. Increasing the frequency of the incident beam, keeping the number of incident photons fixed (this would result in a proportionate increase in energy) increases the maximum kinetic energy of the photoelectrons emitted. Thus the stopping voltage increases. The number of electrons also changes because the probability that each photon results in an emitted electron is a function of photon energy. If the intensity of the incident radiation of a given frequency is increased, there is no effect on the kinetic energy of each photoelectron.

Above the threshold frequency, the maximum kinetic energy of the emitted photoelectron depends on the frequency of the incident light, but is independent of the intensity of the incident light so long as the latter is not too high.

For a given metal and frequency of incident radiation, the rate at which photoelectrons are ejected is directly proportional to the intensity of the incident light. An increase in the intensity of the incident beam (keeping the frequency fixed) increases the magnitude of the photoelectric current, although the stopping voltage remains the same.

The time lag between the incidence of radiation and the emission of a photoelectron is very small, less than 10^{-9} second.

The direction of distribution of emitted electrons peaks in the direction of polarization (the direction of the electric field) of the incident light, if it is linearly polarized.

Mathematical Description

The maximum kinetic energy K_{max} of an ejected electron is given by

$$K_{max} = hf - \varphi,$$

where h is the Planck constant and f is the frequency of the incident photon. The term φ is the work function (sometimes denoted W, or ϕ), which gives the minimum energy required to remove a delocalised electron from the surface of the metal. The work function satisfies

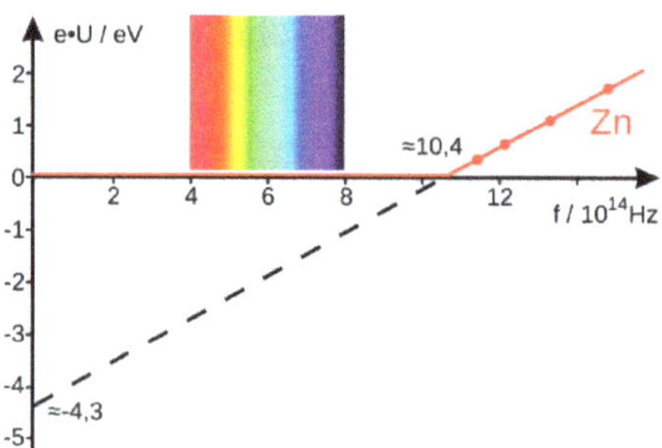

Diagram of the maximum kinetic energy as a function of the frequency of light on zinc

$$\varphi = h f_0,$$

where f_0 is the threshold frequency for the metal. The maximum kinetic energy of an ejected electron is then

$$K_{\max} = h(f - f_0).$$

Kinetic energy is positive, so we must have $f > f_0$ for the photoelectric effect to occur.

Stopping Potential

The relation between current and applied voltage illustrates the nature of the photoelectric effect. For discussion, a light source illuminates a plate P, and another plate electrode Q collects any emitted electrons. We vary the potential between P and Q and measure the current flowing in the external circuit between the two plates.

$$E = \frac{hc}{\lambda}$$

Work function = E.

Cut-off wavelength = λ

Work function and cut off frequency

If the frequency and the intensity of the incident radiation are fixed, the photoelectric current increases gradually with an increase in the positive potential on the collector electrode until all the photoelectrons emitted are collected. The photoelectric current attains a saturation value and does not increase further for any increase in the positive potential. The saturation current increases with the increase of the light intensity. It also increases with greater frequencies due to a greater probability of electron emission when collisions happen with higher energy photons.

If we apply a negative potential to the collector plate Q with respect to the plate P and gradually increase it, the photoelectric current decreases, becoming zero at a certain negative potential. The negative potential on the collector at which the photoelectric current becomes zero is called the *stopping potential* or *cut off* potential

i. For a given frequency of incident radiation, the stopping potential is independent of its intensity.

ii. For a given frequency of incident radiation, the stopping potential is determined by the maximum kinetic energy K_{max} of the photoelectrons that are emitted. If q_e is the charge on the electron and V_0 is the stopping potential, then the work done by the retarding potential in stopping the electron is $q_e V_0$, so we have

$$q_e V_0 = K_{max}.$$

Recalling

$$K_{max} = h(f - f_0),$$

we see that the stopping voltage varies linearly with frequency of light, but depends on the type of material. For any particular material, there is a threshold frequency that must be exceeded, independent of light intensity, to observe any electron emission.

Three-step Model

In the X-ray regime, the photoelectric effect in crystalline material is often decomposed into three steps:

1. Inner photoelectric effect. The hole left behind can give rise to Auger effect, which is visible even when the electron does not leave the material. In molecular solids phonons are excited in this step and may be visible as lines in the final electron energy. The inner photoeffect has to be dipole allowed. The transition rules for atoms translate via the tight-binding model onto the crystal. They are similar in geometry to plasma oscillations in that they have to be transversal.

2. Ballistic transport of half of the electrons to the surface. Some electrons are scattered.

3. Electrons escape from the material at the surface.

In the three-step model, an electron can take multiple paths through these three steps. All paths can interfere in the sense of the path integral formulation. For surface states and molecules the three-step model does still make some sense as even most atoms have multiple electrons which can scatter the one electron leaving.

History

When a surface is exposed to electromagnetic radiation above a certain threshold frequency (typically visible light for alkali metals, near ultraviolet for other metals, and extreme ultraviolet for non-metals), the radiation is absorbed and electrons are emitted. Light, and especially ultra-violet light, discharges negatively electrified bodies with the production of rays of the same nature as cathode rays. Under certain circumstances it can directly ionize gases. The first of these phenomena was discovered by Hertz and Hallwachs in 1887. The second was announced first by Philipp Lenard in 1900.

The ultra-violet light to produce these effects may be obtained from an arc lamp, or by burning magnesium, or by sparking with an induction coil between zinc or cadmium terminals, the light from which is very rich in ultra-violet rays. Sunlight is not rich in ultra-violet rays, as these have been absorbed by the atmosphere, and it does not produce nearly so large an effect as the arc-light. Many substances besides metals discharge negative electricity under the action of ultraviolet light: lists of these substances will be found in papers by G. C. Schmidt and O. Knoblauch.

19th Century

In 1839, Alexandre Edmond Becquerel discovered the photovoltaic effect while studying the effect of light on electrolytic cells. Though not equivalent to the photoelectric effect, his work on photovoltaics was instrumental in showing a strong relationship between light and electronic properties of materials. In 1873, Willoughby Smith discovered photoconductivity in selenium while testing the metal for its high resistance properties in conjunction with his work involving submarine telegraph cables.

Johann Elster (1854–1920) and Hans Geitel (1855–1923), students in Heidelberg, developed the first practical photoelectric cells that could be used to measure the intensity of light. Elster and Geitel had investigated with great success the effects produced by light on electrified bodies.

In 1887, Heinrich Hertz observed the photoelectric effect and the production and reception of electromagnetic waves. He published these observations in the journal Annalen der Physik. His receiver consisted of a coil with a spark gap, where a spark would be seen upon detection of electromagnetic waves. He placed the apparatus in a darkened box to see the spark better. However, he noticed that the maximum spark length was reduced when in the box. A glass panel placed between the source of electromagnetic waves and the receiver absorbed ultraviolet radiation that assisted the electrons in jumping across the gap. When removed, the spark length would increase. He observed no decrease in spark length when he replaced glass with quartz, as quartz does not absorb UV radiation. Hertz concluded his months of investigation and reported the results obtained. He did not further pursue investigation of this effect.

The discovery by Hertz in 1887 that the incidence of ultra-violet light on a spark gap facilitated the passage of the spark, led immediately to a series of investigations by Hallwachs, Hoor, Righi and Stoletow on the effect of light, and especially of ultra-violet light, on charged bodies. It was proved by these investigations that a newly cleaned surface of zinc, if charged with negative electricity, rapidly loses this charge however small it may be when ultra-violet light falls upon the surface; while if the surface is uncharged to begin with, it acquires a positive charge when exposed to the light, the negative electrification going out into the gas by which the metal is surrounded; this positive electrification can be much increased by directing a strong airblast against the surface. If however the zinc surface is positively electrified it suffers no loss of charge when exposed to the light: this result has been questioned, but a very careful examination of the phenomenon by Elster and Geitel has shown that the loss observed under certain circumstances is due to the discharge by the light reflected from the zinc surface of negative electrification on neighbouring conductors induced by the positive charge, the negative electricity under the influence of the electric field moving up to the positively electrified surface.

Heinrich Rudolf Hertz

With regard to the *Hertz effect*, the researches from the start showed a great complexity of the phenomenon of photoelectric fatigue — that is, the progressive diminution of the effect observed upon fresh metallic surfaces. According to an important research by Wilhelm Hallwachs, ozone played an important part in the phenomenon. However, other elements enter such as oxidation, the humidity, the mode of polish of the surface, etc. It was at the time not even sure that the fatigue is absent in a vacuum.

In the period from February 1888 and until 1891, a detailed analysis of photoeffect was performed by Aleksandr Stoletov with results published in 6 works; four of them in *Comptes Rendus*, one review in *Physikalische Revue* (translated from Russian), and the last work in *Journal de Physique*. First, in these works Stoletov invented a new experimental setup which was more suitable for a quantitative analysis of photoeffect. Using this setup, he discovered the direct proportionality between the intensity of light and the induced photo electric current (the first law of photoeffect or Stoletov's law). One of his other findings resulted from measurements of the dependence of the intensity of the electric photo current on the gas pressure, where he found the existence of an optimal gas pressure P_m corresponding to a maximum photocurrent; this property was used for a creation of solar cells.

In 1899, J. J. Thomson investigated ultraviolet light in Crookes tubes. Thomson deduced that the ejected particles were the same as those previously found in the cathode ray, later called electrons, which he called "corpuscles". In the research, Thomson enclosed a metal plate (a cathode) in a vacuum tube, and exposed it to high frequency radiation. It was thought that the oscillating electromagnetic fields caused the atoms' field to resonate and, after reaching a certain amplitude, caused a subatomic "corpuscle" to be emitted, and current to be detected. The amount of this current varied with the intensity and colour of the radiation. Larger radiation intensity or frequency would produce more current.

20th Century

The discovery of the ionization of gases by ultra-violet light was made by Philipp Lenard in 1900. As the effect was produced across several centimeters of air and made very great positive and small negative ions, it was natural to interpret the phenomenon, as did J. J. Thomson, as a *Hertz effect* upon the solid or liquid particles present in the gas.

German physicist Philipp Lenard

In 1902, Lenard observed that the energy of individual emitted electrons increased with the frequency (which is related to the color) of the light.

This appeared to be at odds with Maxwell's wave theory of light, which predicted that the electron energy would be proportional to the intensity of the radiation.

Lenard observed the variation in electron energy with light frequency using a powerful electric arc lamp which enabled him to investigate large changes in intensity, and that had sufficient power to enable him to investigate the variation of potential with light frequency. His experiment directly measured potentials, not electron kinetic energy: he found the electron energy by relating it to the maximum stopping potential (voltage) in a phototube. He found that the calculated maximum electron kinetic energy is determined by the frequency of the light. For example, an increase in frequency results in an increase in the maximum kinetic energy calculated for an electron upon liberation – ultraviolet radiation would require a higher applied stopping potential to stop current in a phototube than blue light. However Lenard's results were qualitative rather than quantitative because of the difficulty in performing the experiments: the experiments needed to be done on freshly cut metal so that the pure metal was observed, but it oxidised in a matter of minutes even in the partial vacuums he used. The current emitted by the surface was determined by the light's intensity, or brightness: doubling the intensity of the light doubled the number of electrons emitted from the surface.

The researches of Langevin and those of Eugene Bloch have shown that the greater part of the Lenard effect is certainly due to this 'Hertz effect'. The Lenard effect upon the gas itself nevertheless does exist. Refound by J. J. Thomson and then more decisively by Frederic Palmer, Jr., it was studied and showed very different characteristics than those at first attributed to it by Lenard.

Einstein, in 1905, when he wrote the *Annus Mirabilis* papers

In 1905, Albert Einstein solved this apparent paradox by describing light as composed of discrete quanta, now called photons, rather than continuous waves. Based upon Max Planck's theory of black-body radiation, Einstein theorized that the energy in each quantum of light was equal to the frequency multiplied by a constant, later called Planck's constant. A photon above a threshold frequency has the required energy to eject a single electron, creating the observed effect. This discovery led to the quantum revolution in physics and earned Einstein the Nobel Prize in Physics in 1921. By wave-particle duality the effect can be analyzed purely in terms of waves though not as conveniently.

Albert Einstein's mathematical description of how the photoelectric effect was caused by absorption of quanta of light was in one of his 1905 papers, named "*On a Heuristic Viewpoint Concerning the Production and Transformation of Light*". This paper proposed the simple description of "light quanta", or photons, and showed how they explained such phenomena as the photoelectric effect. His simple explanation in terms of absorption of discrete quanta of light explained the features of the phenomenon and the characteristic frequency.

The idea of light quanta began with Max Planck's published law of black-body radiation ("*On the Law of Distribution of Energy in the Normal Spectrum*") by assuming that

Hertzian oscillators could only exist at energies E proportional to the frequency f of the oscillator by $E = hf$, where h is Planck's constant. By assuming that light actually consisted of discrete energy packets, Einstein wrote an equation for the photoelectric effect that agreed with experimental results. It explained why the energy of photoelectrons was dependent only on the *frequency* of the incident light and not on its *intensity*: a low-intensity, high-frequency source could supply a few high energy photons, whereas a high-intensity, low-frequency source would supply no photons of sufficient individual energy to dislodge any electrons. This was an enormous theoretical leap, but the concept was strongly resisted at first because it contradicted the wave theory of light that followed naturally from James Clerk Maxwell's equations for electromagnetic behavior, and more generally, the assumption of infinite divisibility of energy in physical systems. Even after experiments showed that Einstein's equations for the photoelectric effect were accurate, resistance to the idea of photons continued, since it appeared to contradict Maxwell's equations, which were well-understood and verified.

Robert Millikan (picture around 1923), who first experimentally showed Einstein's prediction about the photoelectric effect was correct.

Einstein's work predicted that the energy of individual ejected electrons increases linearly with the frequency of the light. Perhaps surprisingly, the precise relationship had not at that time been tested. By 1905 it was known that the energy of photoelectrons increases with increasing *frequency* of incident light and is independent of the *intensity* of the light. However, the manner of the increase was not experimentally determined until 1914 when Robert Andrews Millikan showed that Einstein's prediction was correct.

The photoelectric effect helped to propel the then-emerging concept of wave–particle duality in the nature of light. Light simultaneously possesses the characteristics of both waves and particles, each being manifested according to the circumstances. The effect was impossible to understand in terms of the classical wave description of light, as the energy of the emitted electrons did not depend on the intensity of the incident radiation. Classical theory predicted that the electrons would 'gather up' energy over a period of time, and then be emitted.

Uses and Effects

Photomultipliers

These are extremely light-sensitive vacuum tubes with a photocathode coated onto part (an end or side) of the inside of the envelope. The photocathode contains combinations of materials such as caesium, rubidium and antimony specially selected to provide a low work function, so when illuminated even by very low levels of light, the photocathode readily releases electrons. By means of a series of electrodes (dynodes) at ever-higher potentials, these electrons are accelerated and substantially increased in number through secondary emission to provide a readily detectable output current. Photomultipliers are still commonly used wherever low levels of light must be detected.

Image Sensors

Video camera tubes in the early days of television used the photoelectric effect, for example, Philo Farnsworth's "Image dissector" used a screen charged by the photoelectric effect to transform an optical image into a scanned electronic signal.

Gold-leaf Electroscope

Gold-leaf electroscopes are designed to detect static electricity. Charge placed on the metal cap spreads to the stem and the gold leaf of the electroscope. Because they then have the same charge, the stem and leaf repel each other. This will cause the leaf to bend away from the stem.

The electroscope is an important tool in illustrating the photoelectric effect. For example, if the electroscope is negatively charged throughout, there is an excess of electrons and the leaf is separated from the stem. If high-frequency light shines on the cap, the electroscope discharges and the leaf will fall limp. This is because the frequency of the light shining on the cap is above the cap's threshold frequency. The photons in the light have enough energy to liberate electrons from the cap, reducing its negative charge. This will discharge a negatively charged electroscope and further charge a positive electroscope. However, if the electromagnetic radiation hitting the metal cap does not have a high enough frequency (its frequency is below the threshold value for the cap), then the leaf will never discharge, no matter how long one shines the low-frequency light at the cap.

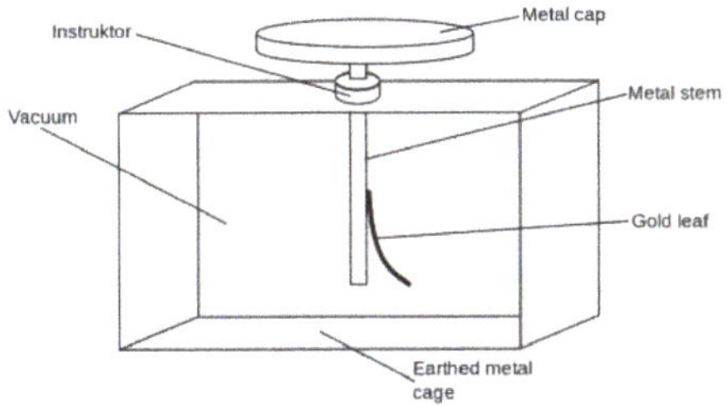

The gold leaf electroscope

Photoelectron Spectroscopy

Since the energy of the photoelectrons emitted is exactly the energy of the incident photon minus the material's work function or binding energy, the work function of a sample can be determined by bombarding it with a monochromatic X-ray source or UV source, and measuring the kinetic energy distribution of the electrons emitted.

Photoelectron spectroscopy is usually done in a high-vacuum environment, since the electrons would be scattered by gas molecules if they were present. However, some companies are now selling products that allow photoemission in air. The light source can be a laser, a discharge tube, or a synchrotron radiation source.

The concentric hemispherical analyser (CHA) is a typical electron energy analyzer, and uses an electric field to change the directions of incident electrons, depending on their kinetic energies. For every element and core (atomic orbital) there will be a different binding energy. The many electrons created from each of these combinations will show up as spikes in the analyzer output, and these can be used to determine the elemental composition of the sample.

Spacecraft

The photoelectric effect will cause spacecraft exposed to sunlight to develop a positive charge. This can be a major problem, as other parts of the spacecraft in shadow develop a negative charge from nearby plasma, and the imbalance can discharge through delicate electrical components. The static charge created by the photoelectric effect is self-limiting, though, because a more highly charged object gives up its electrons less easily.

Moon Dust

Light from the sun hitting lunar dust causes it to become charged through the photoelectric effect. The charged dust then repels itself and lifts off the surface of the Moon by electrostatic levitation. This manifests itself almost like an "atmosphere of dust", visible as a thin haze and blurring of distant features, and visible as a dim glow after the sun has set. This was first photographed by the Surveyor program probes in the 1960s. It is thought that the smallest particles are repelled up to kilometers high, and that the particles move in "fountains" as they charge and discharge.

Night Vision Devices

Photons hitting a thin film of alkali metal or semiconductor material such as gallium arsenide in an image intensifier tube cause the ejection of photoelectrons due to the photoelectric effect. These are accelerated by an electrostatic field where they strike a phosphor coated screen, converting the electrons back into photons. Intensification of the signal is achieved either through acceleration of the electrons or by increasing the number of electrons through secondary emissions, such as with a micro-channel plate. Sometimes a combination of both methods is used. Additional kinetic energy is required to move an electron out of the conduction band and into the vacuum level. This is known as the electron affinity of the photocathode and is another barrier to photoemission other than the forbidden band, explained by the band gap model. Some materials such as Gallium Arsenide have an effective electron affinity that is below the level of the conduction band. In these materials, electrons that move to the conduction band are all of sufficient energy to be emitted from the material and as such, the film that absorbs photons can be quite thick. These materials are known as negative electron affinity materials.

Cross Section

The photoelectric effect is one interaction mechanism between photons and atoms. It is one of 12 theoretically possible interactions.

At the high photon energies comparable to the electron rest energy of 511 keV, Compton scattering, another process, may take place. Above twice this (1.022 MeV) pair production may take place. Compton scattering and pair production are examples of two other competing mechanisms.

Indeed, even if the photoelectric effect is the favoured reaction for a particular single-photon bound-electron interaction, the result is also subject to statistical processes and is not guaranteed, albeit the photon has certainly disappeared and a bound electron has been excited (usually K or L shell electrons at gamma ray energies). The probability of the photoelectric effect occurring is measured by the cross section of interaction, σ. This has been found to be a function of the atomic number of the target atom and photon energy. A crude approximation, for photon energies above the highest atomic binding energy, is given by:

$$\sigma = \text{constant} \cdot \frac{Z^n}{E^3}$$

Here Z is atomic number and n is a number which varies between 4 and 5. (At lower photon energies a characteristic structure with edges appears, K edge, L edges, M edges, etc.) The obvious interpretation follows that the photoelectric effect rapidly decreases in significance, in the gamma ray region of the spectrum, with increasing photon energy, and that photoelectric effect increases steeply with atomic number. The corollary is that high-Z materials make good gamma-ray shields, which is the principal reason that lead ($Z = 82$) is a preferred and ubiquitous gamma radiation shield.

References

- Lai, Shu T. (2011). Fundamentals of Spacecraft Charging: Spacecraft Interactions with Space Plasmas (illustrated ed.). Princeton University Press. pp. 1–6. ISBN 978-0-691-12947-1.
- Cox, James F. (2001). Fundamentals of linear electronics: integrated and discrete. Cengage Learning. pp. 91–. ISBN 978-0-7668-3018-9.

- Tavernier, Filip and Steyaert, Michiel (2011) High-Speed Optical Receivers with Integrated Photodiode in Nanoscale CMOS. Springer. ISBN 1-4419-9924-8.
- Riordan, Michael and Hoddeson, Lillian (1998). Crystal Fire: The Invention of the Transistor and the Birth of the Information Age. ISBN 9780393318517.
- Held. G, Introduction to Light Emitting Diode Technology and Applications, CRC Press, (Worldwide, 2008). Ch. 5 p. 116. ISBN 1-4200-7662-0
- Brooker, Graham (2009) Introduction to Sensors for Ranging and Imaging, ScitTech Publishing. p. 87. ISBN 9781891121746
- Knoll, F.G. (2010). Radiation detection and measurement, 4th ed. Wiley, Hoboken, NJ. p. 298. ISBN 978-0-470-13148-0
- Gao, Wei (2010). Precision Nanometrology: Sensors and Measuring Systems for Nanomanufacturing. Springer. pp. 15–16. ISBN 978-1-84996-253-7.
- Crecraft, David; Stephen Gergely (2002). Analog Electronics: Circuits, Systems and Signal Processing. Butterworth-Heinemann. p. 110. ISBN 0-7506-5095-8.
- Horowitz, Paul; Winfield Hill (1989). The Art of Electronics, 2nd Ed. London: Cambridge University Press. p. 44. ISBN 0-521-37095-7.
- Sung-Mo Kang and Yusuf Leblebici (2002). CMOS Digital Integrated Circuits Analysis & Design. McGraw–Hill Professional. ISBN 0-07-246053-9.
- Sears, F. W.; Zemansky, M. W.; Young, H. D. (1983). University Physics (6th ed.). Addison-Wesley. pp. 843–844. ISBN 0-201-07195-9.
- Mee, C.; Crundell, M.; Arnold, B.; Brown, W. (2011). International A/AS Level Physics. Hodder Education. p. 241. ISBN 978-0-340-94564-3.
- Fromhold, A. T. (1991). Quantum Mechanics for Applied Physics and Engineering. Courier Dover Publications. pp. 5–6. ISBN 978-0-486-66741-6.
- Robert Bud; Deborah Jean Warner (1998). Instruments of Science: An Historical Encyclopedia. Science Museum, London, and National Museum of American History, Smithsonian Institution. ISBN 978-0-8153-1561-2.
- Thomson, J. J. (2005). Conduction of Electricity Through Gases. Watchmaker Publishing. ISBN 978-1-929148-49-3. Retrieved 9 July 2011.
- Buchwald, Jed; Warwick, Andrew, eds. (2004). Histories of the Electron: The Birth of Microphysics (PDF) (illustrated, reprint ed.). MIT Press. pp. 21–23. ISBN 978-0-262-52424-7.
- Knight, Randall D. (2004) Physics for Scientists and Engineers With Modern Physics: A Strategic Approach, Pearson-Addison-Wesley, p. 1224, ISBN 0-8053-8685-8.
- Penrose, Roger (2005) The Road to Reality: A Complete Guide to the Laws of the Universe, Knopf, p. 502, ISBN 0-679-45443-8
- Timothy, J. Gethyn (2010) in Huber, Martin C.E. (ed.) Observing Photons in Space, ISSI Scientific Report 009, ESA Communications, pp. 365–408, ISBN 978-92-9221-938-3
- Tsokos, K. A. (2010). Cambridge Physics for the IB Diploma (revised ed.). Cambridge University Press. ISBN 978-0-521-13821-5.

4

Photovoltaics: A Comprehensive Study

The following text is an overview of the subject matter incorporating all the major aspects of photovoltaics. Photovoltaic cells convert light into electricity with the help of semiconducting materials such as silicon and selenium. This chapter also focuses on topics such as photovoltaic system, grid-connected photovoltaic power system, building-integrated photovoltaics and concentrator photovoltaics.

Photovoltaics

Photovoltaics (PV) covers the conversion of light into electricity using semiconducting materials that exhibit the photovoltaic effect, a phenomenon studied in physics, photochemistry, and electrochemistry.

The Solar Settlement, a sustainable housing community project in Freiburg, Germany.

A typical photovoltaic system employs solar panels, each comprising a number of solar cells, which generate electrical power. The first step is the photoelectric effect followed by an electrochemical process where crystallized atoms, ionized in a series, generate an electric current. PV Installations may be ground-mounted, rooftop mounted or wall mounted.

Solar PV generates no pollution. The direct conversion of sunlight to electricity occurs without any moving parts. Photovoltaic systems have been used for fifty years in specialized applications, standalone and grid-connected PV systems have been in use for more than twenty years. They were first mass-produced in 2000, when German environmentalists and the Eurosolar organization got government funding for a ten thousand roof program.

On the other hand, grid-connected PV systems have the major disadvantage that the power output is dependent on direct sunlight, so about 10-25% is lost if a tracking system is not used, since the cell wil not be directly facing the sun at all times.Power output is also adversely affected by weather conditions such as the amount of dust and water vapour in the air or the amount of cloud cover. This means that, in the national grid for example, this power has to be made up by other power sources: hydrocarbon, nuclear, hydroelectric or wind energy.

Double glass photovoltaic solar modules, installed in a support structure.

Advances in technology and increased manufacturing scale have reduced the cost, increased the reliability, and increased the efficiency of photovoltaic instalations and the levelised cost of electricity from PV is competitive, on a kilowatt/ hour basis, with conventional electricity sources in an expanding list of geographic regions. Solar PV regularly costs USD 0.05-0.10 per kilowatt-hour (kWh) in Europe, China, India, South Africa and the United States. In 2015, record low prices were set in the United Arab Emirates (5.84 cents/kWh), Peru (4.8 cents/kWh) and Mexico (4.8 cents/kWh). In May 2016, a solar PV auction in Dubai attracted a bid of 3 cents/kWh.

Solar panels on the International Space Station

Net metering and financial incentives, such as preferential feed-in tariffs for solar-generated electricity, have supported solar PV installations in many countries. More than 100 countries now use solar PV. After hydro and wind powers, PV is the third renewable energy source in terms of globally capacity. In 2014, worldwide installed PV capacity increased to 177 gigawatts (GW), which is two percent of global electricity demand.

China, followed by Japan and the United States, is the fastest growing market, while Germany remains the world's largest producer (both in per capita and absolute terms), with solar PV providing seven percent of annual domestic electricity consumption.

With current technology (as of 2013), photovoltaics recoups the energy needed to manufacture them in 1.5 years in Southern Europe and 2.5 years in Northern Europe.

Solar Cells

Solar cells generate electricity directly from sunlight.

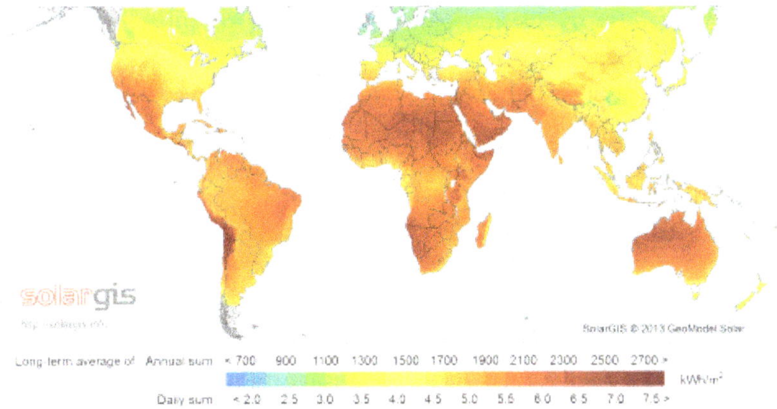

Average insolation. Note that this is for a horizontal surface. Solar panels are normally propped up at an angle and receive more energy per unit area.

Photovoltaics are best known as a method for generating electric power by using solar cells to convert energy from the sun into a flow of electrons. The photovoltaic effect refers to photons of light exciting electrons into a higher state of energy, allowing them to act as charge carriers for an electric current. The photovoltaic effect was first observed by Alexandre-Edmond Becquerel in 1839. The term photovoltaic denotes the unbiased operating mode of a photodiode in which current through the device is entirely due to the transduced light energy. Virtually all photovoltaic devices are some type of photodiode.

Solar cells produce direct current electricity from sun light which can be used to power equipment or to recharge a battery. The first practical application of photovoltaics was to power orbiting satellites and other spacecraft, but today the majority of photovoltaic modules are used for grid connected power generation. In this case an inverter is required to convert the DC to AC. There is a smaller market for off-grid power for remote dwellings, boats, recreational vehicles, electric cars, roadside emergency telephones, remote sensing, and cathodic protection of pipelines.

Photovoltaic power generation employs solar panels composed of a number of solar cells containing a photovoltaic material. Materials presently used for photovoltaics include monocrystalline silicon, polycrystalline silicon, amorphous silicon, cadmium telluride, and copper indium gallium selenide/sulfide. Copper solar cables connect modules (module cable), arrays (array cable), and sub-fields. Because of the growing demand for renewable energy sources, the manufacturing of solar cells and photovoltaic arrays has advanced considerably in recent years.

Solar photovoltaics power generation has long been seen as a clean energy technology which draws upon the planet's most plentiful and widely distributed renewable energy source – the sun. The technology is "inherently elegant" in that the direct conversion of sunlight to electricity occurs without any moving parts or environmental emissions during operation. It is well proven, as photovoltaic systems have now been used for fifty years in specialised applications, and grid-connected systems have been in use for over twenty years.

Cells require protection from the environment and are usually packaged tightly behind a glass sheet. When more power is required than a single cell can deliver, cells are electrically connected together to form photovoltaic modules, or solar panels. A single module is enough to power an emergency telephone, but for a house or a power plant the modules must be arranged in multiples as arrays.

Photovoltaic power capacity is measured as maximum power output under standardized test conditions (STC) in "W_p" (watts peak). The actual power output at a particular point in time may be less than or greater than this standardized, or "rated," value, depending on geographical location, time of day, weather conditions, and other factors. Solar photovoltaic array capacity factors are typically under 25%, which is lower than many other industrial sources of electricity.

Current Developments

For best performance, terrestrial PV systems aim to maximize the time they face the sun. Solar trackers achieve this by moving PV panels to follow the sun. The increase can be by as much as 20% in winter and by as much as 50% in summer. Static mounted systems can be optimized by analysis of the sun path. Panels are often set to latitude tilt, an angle equal to the latitude, but performance can be improved by adjusting the angle for summer or winter. Generally, as with other semiconductor devices, temperatures above room temperature reduce the performance of photovoltaics.

A number of solar panels may also be mounted vertically above each other in a tower, if the zenith distance of the Sun is greater than zero, and the tower can be turned horizontally as a whole and each panels additionally around a horizontal axis. In such a tower the panels can follow the Sun exactly. Such a device may be described as a ladder mounted on a turnable disk. Each step of that ladder is the middle axis of a rectangular solar panel. In case the zenith distance of the Sun reaches zero, the "ladder" may be rotated to the north or the south to avoid a solar panel producing a shadow on a lower solar panel. Instead of an exactly vertical tower one can choose a tower with an axis directed to the polar star, meaning that it is parallel to the rotation axis of the Earth. In this case the angle between the axis and the Sun is always larger than 66 degrees. During a day it is only necessary to turn the panels around this axis to follow the Sun. Installations may be ground-mounted (and sometimes integrated with farming and grazing) or built into the roof or walls of a building (building-integrated photovoltaics).

Another recent development involves the makeup of solar cells. Perovskite is a very inexpensive material which is being used to replace the expensive crystalline silicon which is still part of a standard PV cell build to this day. Michael Graetzel, Director of the Laboratory of Photonics and Interfaces at EPFL says, "Today, efficiency has peaked at 18 percent, but it's expected to get even higher in the future." This is a significant claim, as 20% efficiency is typical among solar panels which use more expensive materials.

Efficiency

Electrical efficiency (also called conversion efficiency) is a contributing factor in the selection of a photovoltaic system. However, the most efficient solar panels are typically the most expensive, and may not be commercially available. Therefore, selection is also driven by cost efficiency and other factors.

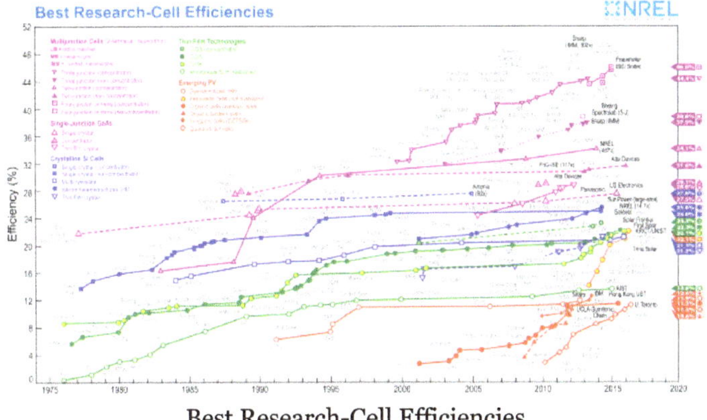

Best Research-Cell Efficiencies

The electrical efficiency of a PV cell is a physical property which represents how much electrical power a cell can produce for a given insolation. The basic expression for maximum efficiency of a photovoltaic cell is given by the ratio of output power to the incident solar power (radiation flux times area)

$$\eta = \frac{P_{max}}{E \cdot A_{cell}}.$$

The efficiency is measured under ideal laboratory conditions and represents the maximum achievable efficiency of the PV material. Actual efficiency is influenced by the output Voltage, current, junction temperature, light intensity and spectrum.

The most efficient type of solar cell to date is a multi-junction concentrator solar cell with an efficiency of 46.0% produced by Fraunhofer ISE in December 2014. The highest efficiencies achieved without concentration include a material by Sharp Corporation at 35.8% using a proprietary triple-junction manufacturing technology in 2009, and Boeing Spectrolab (40.7% also using a triple-layer design). The US company SunPower produces cells that have an efficiency of 21.5%, well above the market average of 12–18%.

There is an ongoing effort to increase the conversion efficiency of PV cells and modules, primarily for competitive advantage. In order to increase the efficiency of solar cells, it is important to choose a semiconductor material with an appropriate band gap that matches the solar spectrum. This will enhance the electrical and optical properties. Improving the method of charge collection is also useful for increasing the efficiency. There are several groups of materials that are being developed. Ultrahigh-efficiency devices (η>30%) are made by using GaAs and GaInP2 semiconductors with multijunction tandem cells. High-quality, single-crystal silicon materials are used to achieve high-efficiency, low cost cells (η>20%).

Recent developments in Organic photovoltaic cells (OPVs) have made significant advancements in power conversion efficiency from 3% to over 15% since their introduction in the 1980s. To date, the highest reported power conversion efficiency ranges from 6.7% to 8.94% for small molecule, 8.4%–10.6% for polymer OPVs, and 7% to 21% for perovskite OPVs. OPVs are expected to play a major role in the PV market. Recent improvements have increased the efficiency and lowered cost, while remaining environmentally-benign and renewable.

Several companies have begun embedding power optimizers into PV modules called smart modules. These modules perform maximum power point tracking (MPPT) for each module individually, measure performance data for monitoring, and provide additional safety features. Such modules can also compensate for shading effects, wherein a shadow falling across a section of a module causes the electrical output of one or more strings of cells in the module to decrease.

One of the major causes for the decreased performance of cells is overheating. The efficiency of a solar cell declines by about 0.5% for every 1 degree Celsius increase in temperature. This means that a 100 degree increase in surface temperature could decrease the efficiency of a solar cell by about half. Self-cooling solar cells are one solution to this problem. Rather than using energy to cool the surface, pyramid and cone shapes can be formed from silica, and attached to the surface of a solar panel. Doing so allows visible light to reach the solar cells, but reflects infrared rays (which carry heat).

Growth

Solar photovoltaics is growing rapidly and worldwide installed capacity reached at least 177 gigawatts (GW) by the end of 2014. The total power output of the world's PV capacity in a calendar year is now beyond 200 TWh of electricity. This represents 1% of worldwide electricity demand. More than 100 countries use solar PV. China, followed by Japan and the United States is now the fastest growing market, while Germany remains the world's largest producer, contributing more than 7% to its national electricity demands. Photovoltaics is now, after hydro and wind power, the third most important renewable energy source in terms of globally installed capacity.

Several market research and financial companies foresee record-breaking global installation of more than 50 GW in 2015. China is predicted to take the lead from Germany and to become the world's largest producer of PV power by installing another targeted 17.8 GW in 2015. India is expected to install 1.8 GW, doubling its annual installations. By 2018, worldwide photovoltaic capacity is projected to doubled or even triple to 430 GW. Solar Power Europe (formerly known as EPIA) also estimates that photovoltaics will meet 10% to 15% of Europe's energy demand in 2030.

The EPIA/Greenpeace Solar Generation Paradigm Shift Scenario (formerly called Advanced Scenario) from 2010 shows that by the year 2030, 1,845 GW of PV systems could be generating approximately 2,646 TWh/year of electricity around the world. Combined with energy use efficiency improvements, this would represent the electricity needs of more than 9% of the world's population. By 2050, over 20% of all electricity could be provided by photovoltaics.

Michael Liebreich, from Bloomberg New Energy Finance, anticipates a tipping point for solar energy. The costs of power from wind and solar are already below those of conventional electricity generation in some parts of the world, as they have fallen sharply and will continue to do so. He also asserts, that the electrical grid has been greatly expanded worldwide, and is ready to receive and distribute electricity from renewable sources. In addition, worldwide electricity prices came under strong pressure from renewable energy sources, that are, in part, enthusiastically embraced by consumers.

Deutsche Bank sees a "second gold rush" for the photovoltaic industry to come. Grid parity has already been reached in at least 19 markets by January 2014. Photovoltaics will prevail beyond feed-in tariffs, becoming more competitive as deployment increases and prices continue to fall.

In June 2014 Barclays downgraded bonds of U.S. utility companies. Barclays expects more competition by a growing self-consumption due to a combination of decentralized PV-systems and residential electricity storage. This could fundamentally change the utility's business model and transform the system over the next ten years, as prices for these systems are predicted to fall.

Environmental Impacts of Photovoltaic Technologies

Types of Impacts

While solar photovoltaic (PV) cells are promising for clean energy production, their deployment is hindered by production costs, material availability, and toxicity. Life cycle assessment (LCA) is one method of determining environmental impacts from PV. Many studies have been done on the various types of PV including first generation, second generation, and third generation. Usually these PV LCA studies select a cradle to gate system boundary because often at the time the studies are conducted, it is a new technology not commercially available yet and their required balance of system components and disposal methods are unknown.

A traditional LCA can look at many different impact categories ranging from global warming potential, eco-toxicity, human toxicity, water depletion, and many others. Most LCAs of PV have focused on two categories: carbon dioxide equivalents per kWh and energy pay-back time (EPBT). The EPBT is defined as " the time needed to compensate for the total renewable- and non-renewable- primary energy required during the life cycle of a PV system". A 2015 review of EPBT from first and second generation PV suggested that there was greater variation in embedded energy than in efficiency of the cells implying that it was mainly the embedded energy that needs to reduce to have a greater reduction in EPBT. One difficulty in determining impacts due to PV is to determine if the wastes are released to the air, water, or soil during the manufacturing phase. Research is underway to try to understand emissions and releases during the lifetime of PV systems.

Impacts from First-generation PV

Crystalline silicon modules are the most extensively studied PV type in terms of LCA since they are the most commonly used. Mono-crystalline silicon photovoltaic systems (mono-si) have an average efficiency of 14.0%. The cells tend to follow a structure of front electrode, anti-reflection film, n-layer, p-layer, and back electrode, with the sun hitting the front electrode. EPBT ranges from 1.7 to 2.7 years. The cradle to gate of CO_2-eq/kWh ranges from 37.3 to 72.2 grams.

Techniques to produce multi-crystalline silicon (multi-si) photovoltaic cells are simpler and cheaper than mono-si, however tend to make less efficient cells, an average of 13.2%. EPBT ranges from 1.5 to 2.6 years. The cradle to gate of CO_2-eq/kWh ranges from 28.5 to 69 grams. Some studies have looked beyond EPBT and GWP to other environmental impacts. In one such study, conventional energy mix in Greece was compared to multi-si PV and found a 95% overall reduction in impacts including carcinogens, eco-toxicity, acidification, eutrophication, and eleven others.

Impacts from Second Generation

Cadmium telluride (CdTe) is one of the fastest-growing thin film based solar cells which are collectively known as second generation devices. This new thin film device also shares similar performance restrictions (Shockley-Queisser efficiency limit) as conventional Si devices but promises to lower the cost of each device by both reducing material and energy consumption during manufacturing. Today the global market share of CdTe is 5.4%, up from 4.7% in 2008. This technology's highest power conversion efficiency is 21%. The cell structure includes glass substrate (around 2 mm), transparent conductor layer, CdS buffer layer (50–150 nm), CdTe absorber and a metal contact layer.

CdTe PV systems require less energy input in their production than other commercial PV systems per unit electricity production. The average CO_2-eq/kWh is around 18 grams (cradle to gate). CdTe has the fastest EPBT of all commercial PV technologies, which varies between 0.3 and 1.2 years.

Copper Indium Gallium Diselenide (CIGS) is a thin film solar cell based on the copper indium diselenide (CIS) family of chalcopyrite semiconductors. CIS and CIGS are often used interchangeably within the CIS/CIGS community. The cell structure includes soda lime glass as the substrate, Mo layer as the back contact, CIS/CIGS as the absorber layer, cadmium sulfide (CdS) or Zn (S,OH)x as the buffer layer, and ZnO:Al as the front contact. CIGS is approximately 1/100th the thickness of conventional silicon solar cell technologies. Materials necessary for assembly are readily available, and are less costly per watt of solar cell. CIGS based solar devices resist performance degradation over time and are highly stable in the field.

Reported global warming potential impacts of CIGS range from 20.5 – 58.8 grams CO_2-eq/kWh of electricity generated for different solar irradiation (1,700 to 2,200 kWh/m^2/y) and power conversion efficiency (7.8 – 9.12%). EPBT ranges from 0.2 to 1.4 years, while harmonized value of EPBT was found 1.393 years. Toxicity is an issue within the buffer layer of CIGS modules because it contains cadmium and gallium. CIS modules do not contain any heavy metals.

Impacts from Third Generation

Third-generation PVs are designed to combine the advantages of both the first and second generation devices and they do not have Shockley-Queisser efficiency limit, a theoretical limit for first and second generation PV cells. The thickness of a third generation device is less than 1 μm.

One emerging alternative and promising technology is based on an organic-inorganic hybrid solar cell made of methylammonium lead halide perovskites. Perovskite PV cells have progressed rapidly over the past few years and have become one of the most attractive areas for PV research. The cell structure includes a metal back contact (which can be made of Al, Au or Ag), a hole transfer layer (spiro-MeOTAD, P3HT, PTAA, CuSCN, CuI, or NiO), and absorber layer ($CH_3NH_3PbIxBr_3$-x, $CH_3NH_3PbIxCl_3$-x or $CH_3NH_3PbI_3$), an electron transport layer (TiO, ZnO, Al_2O_3 or SnO_2) and a top contact layer (fluorine doped tin oxide or tin doped indium oxide).

There are a limited number of published studies to address the environmental impacts of perovskite solar cells. The major environmental concern is the lead used in the absorber layer. Due to the instability of perovskite cells lead may eventually be exposed to fresh water during the use phase. These LCA studies looked at human and ecotoxicity of perovskite solar cells and found they were surprisingly low and may not be an environmental issue. Global warming potential of perovskite PVs were found to be in the range of 24–1500 grams CO_2-eq/kWh electricity production. Similarly, reported EPBT of the published paper range from 0.2 to 15 years. The large range of reported values highlight the uncertainties associated with these studies. Celik et al. (2016) critically discussed the assumptions made in perovskite PV LCA studies.

Two new promising thin film technologies are copper zinc tin sulfide (Cu_2ZnSnS_4 or CZTS) and zinc phosphide (Zn_3P_2). Both of these thin films are currently only produced in the lab but may be commercialized in the future. Their manufacturing processes are expected to be similar to those of current thin film technologies of CIGS and CdTe, respectively. Yet, contrary to CIGS and CdTe, CZTS and Zn_3P_2 are made from earth abundant, nontoxic materials and have the potential to produce more electricity annually than the current worldwide consumption. While CZTS and Zn_3P_2 offer good promise for these reasons, the specific environmental implications of their commercial production are not yet known. Global warming potential of CZTS and Zn_3P_2 were found 38 and 30 grams CO_2-eq/kWh while their corresponding EPBT were found 1.85 and 0.78 years, respectively. Overall, CdTe and Zn_3P_2 have similar environmental impacts but can slightly outperform CIGS and CZTS.

Organic and polymer photovoltaic (OPV) are a relatively new area of research. The tradition OPV cell structure layers consist of a semi-transparent electrode, electron blocking layer, tunnel junction, holes blocking layer, electrode, with the sun hitting the transparent electrode. OPV replaces silver with carbon as an electrode material lowering manufacturing cost and making them more environmentally friendly. OPV are flexible, low weight, and work well with roll-to roll manufacturing for mass production. OPV uses "only abundant elements coupled to an extremely low embodied energy through very low processing temperatures using only ambient processing conditions on simple printing equipment enabling energy pay-back times". Current efficiencies range from 1–6.5%, however theoretical analyses show promise beyond 10% efficiency.

Many different configurations of OPV exist using different materials for each layer. OPV technology rivals existing PV technologies in terms of EPBT even if they currently present a shorter operational lifetime. A 2013 study analyzed 12 different configurations all with 2% efficiency, the EPBT ranged from 0.29–0.52 years for 1 m^2 of PV. The average CO_2-eq/kWh for OPV is 54.922 grams.

Economics

There have been major changes in the underlying costs, industry structure and market prices of solar photovoltaics technology, over the years, and gaining a coherent picture of the shifts occurring across the industry value chain globally is a challenge. This is due to: "the rapidity of cost and price changes, the complexity of the PV supply chain, which involves a large number of manufacturing processes, the balance of system (BOS) and installation costs associated with complete PV systems, the choice of different distribution channels, and differences between regional markets within which PV is being deployed". Further complexities result from the many different policy support initiatives that have been put in place to facilitate photovoltaics commercialisation in various countries.

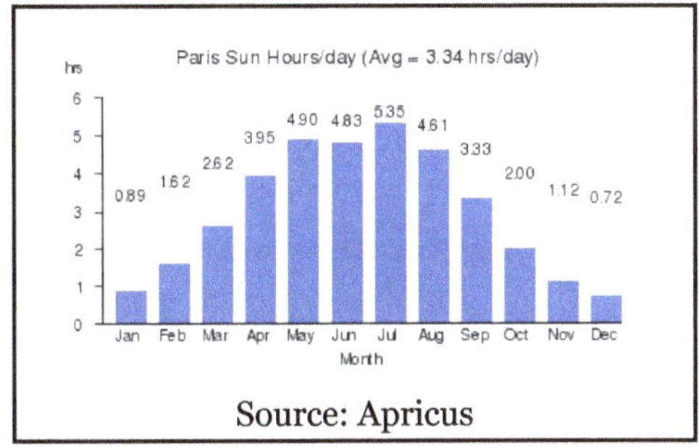

Source: Apricus

The PV industry has seen dramatic drops in module prices since 2008. In late 2011, factory-gate prices for crystalline-silicon photovoltaic modules dropped below the \$1.00/W mark. The \$1.00/W installed cost, is often regarded in the PV industry as marking the achievement of grid parity for PV. Technological advancements, manufacturing process improvements, and industry re-structuring, mean that further price reductions are likely in coming years.

Financial incentives for photovoltaics, such as feed-in tariffs, have often been offered to electricity consumers to install and operate solar-electric generating systems. Government has sometimes also offered incentives in order to encourage the PV industry to achieve the economies of scale needed to compete where the cost of PV-generated electricity is above the cost from the existing grid. Such policies are implemented to promote national or territorial energy independence, high tech job creation and reduction of carbon dioxide emissions which cause global warming. Due to economies of scale solar panels get less costly as people use and buy more—as manufacturers increase production to meet demand, the cost and price is expected to drop in the years to come.

Solar cell efficiencies vary from 6% for amorphous silicon-based solar cells to 44.0% with multiple-junction concentrated photovoltaics. Solar cell energy conversion efficiencies for commercially available photovoltaics are around 14–22%. Concentrated photovoltaics (CPV) may reduce cost by concentrating up to 1,000 suns (through magnifying lens) onto a smaller sized photovoltaic cell. However, such concentrated solar power requires sophisticated heat sink designs, otherwise the photovoltaic cell overheats, which reduces its efficiency and life. To further exacerbate the concentrated cooling design, the heat sink must be passive, otherwise the power required for active cooling would reduce the overall efficiency and economy.

Crystalline silicon solar cell prices have fallen from \$76.67/Watt in 1977 to an estimated \$0.74/Watt in 2013. This is seen as evidence supporting Swanson's law, an observation similar to the famous Moore's Law that states that solar cell prices fall 20% for every doubling of industry capacity.

As of 2011, the price of PV modules has fallen by 60% since the summer of 2008, according to Bloomberg New Energy Finance estimates, putting solar power for the first time on a competitive footing with the retail price of electricity in a number of sunny countries; an alternative and consistent price decline figure of 75% from 2007 to 2012 has also been published, though it is unclear whether these figures are specific to the United States or generally global. The levelised cost of electricity (LCOE) from PV is competitive with conventional electricity sources in an expanding list of geographic regions, particularly when the time of generation is included, as electricity is worth more during the day than at night. There has been fierce competition in the supply chain, and further improvements in the levelised cost of energy for solar lie ahead, posing a growing threat to the dominance of fossil fuel generation sources in the next few years. As time progresses, renewable energy technologies generally get cheaper, while fossil fuels generally get more expensive:

The less solar power costs, the more favorably it compares to conventional power, and the more attractive it becomes to utilities and energy users around the globe. Utility-scale solar power can now be delivered in California at prices well below \$100/MWh (\$0.10/kWh) less than most other peak generators, even those running on low-cost natural gas. Lower solar module costs also stimulate demand from consumer markets where the cost of solar compares very favorably to retail electric rates.

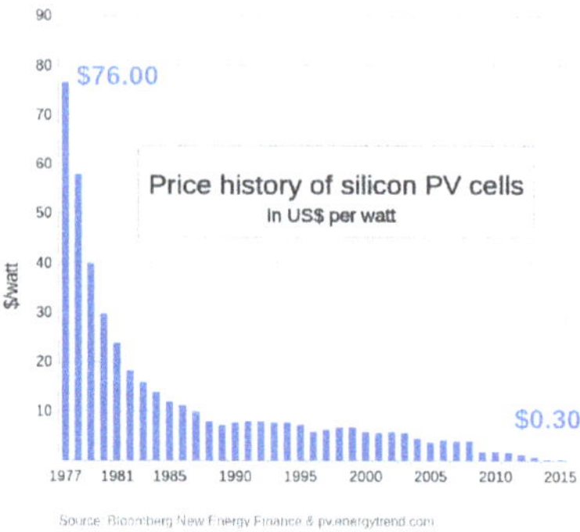

Price per watt history for conventional (c-Si) solar cells since 1977.

As of 2011, the cost of PV has fallen well below that of nuclear power and is set to fall further. The average retail price of solar cells as monitored by the Solarbuzz group fell from $3.50/watt to $2.43/watt over the course of 2011.

For large-scale installations, prices below $1.00/watt were achieved. A module price of 0.60 Euro/watt ($0.78/watt) was published for a large scale 5-year deal in April 2012.

By the end of 2012, the "best in class" module price had dropped to $0.50/watt, and was expected to drop to $0.36/watt by 2017.

In many locations, PV has reached grid parity, which is usually defined as PV production costs at or below retail electricity prices (though often still above the power station prices for coal or gas-fired generation without their distribution and other costs). However, in many countries there is still a need for more access to capital to develop PV projects. To solve this problem securitization has been proposed and used to accelerate development of solar photovoltaic projects. For example, SolarCity offered, the first U.S. asset-backed security in the solar industry in 2013.

Photovoltaic power is also generated during a time of day that is close to peak demand (precedes it) in electricity systems with high use of air conditioning. More generally, it is now evident that, given a carbon price of $50/ton, which would raise the price of coal-fired power by 5c/kWh, solar PV will be cost-competitive in most locations. The declining price of PV has been reflected in rapidly growing installations, totaling about 23 GW in 2011. Although some consolidation is likely in 2012, due to support cuts in the large markets of Germany and Italy, strong growth seems likely to continue for the rest of the decade. Already, by one estimate, total investment in renewables for 2011 exceeded investment in carbon-based electricity generation.

In the case of self consumption payback time is calculated based on how much electricity is not brought from the grid. Additionally, using PV solar power to charge DC batteries, as used in Plug-in Hybrid Electric Vehicles and Electric Vehicles, leads to greater efficiencies. Traditionally, DC generated electricity from solar PV must be converted to AC for buildings, at an average 10% loss during the conversion. An additional efficiency loss occurs in the transition back to DC for battery driven devices and vehicles, and using various interest rates and energy price changes were calculated to find present values that range from $2,057.13 to $8,213.64 (analysis from 2009).

For example, in Germany with electricity prices of 0.25 euro/kWh and Insolation of 900 kWh/kW one kW_p will save 225 euro per year and with installation cost of 1700 euro/kW_p means that the system will pay back in less than 7 years.

Manufacturing

Overall the manufacturing process of creating solar photovoltaics is simple in that it does not require the culmination of many complex or moving parts. Because of the solid state nature of PV systems they often have relatively long lifetimes, anywhere from 10 to 30 years. In order to increase electrical output of a PV system the manufacturer must simply add more photovoltaic components and because of this economies of scale are important for manufacturers as costs decrease with increasing output.

While there are many types of PV systems known to be effective, crystalline silicon PV accounted for around 90% of the worldwide production of PV in 2013. Manufacturing silicon PV systems has several steps. First, polysilicon is processed from mined quartz until it is very pure (semi-conductor grade). This is melted down when small amounts of Boron, a group III element, are added to make a p-type semiconductor rich in electron holes. Typically using a seed crystal, an ingot of this solution is grown from the liquid polycrystalline. The ingot may also be cast in a mold. Wafers of this semiconductor material are cut from the bulk material with wire saws, and then go through surface etching before being cleaned. Next, the wafers are placed into a phosphorus vapor deposition furnace which lays a very thin layer of phosphorus, a group V element, which creates an N-type semiconducting surface. To reduce energy losses an anti-reflective coating is added to the surface, along with electrical contacts. After finishing the cell, cells are connected via electrical circuit according to the specific application and prepared for shipping and installation.

Crystalline silicon photovoltaics are only one type of PV, and while they represent the majority of solar cells produced currently there are many new and promising technologies that have the potential to be scaled up to meet future energy needs.

Another newer technology, thin-film PV, are manufactured by depositing semiconducting layers on substrate in vacuum. The substrate is often glass or stainless-steel, and these semiconducting layers are made of many types of materials including cadmium telluride (CdTe), copper indium diselenide (CIS), copper indium gallium diselenide (CIGS), and amorphous silicon (a-Si). After being deposited onto the substrate the semiconducting layers are separated and connected by electrical circuit by laser-scribing. Thin-film photovoltaics now make up around 20% of the overall production of PV because of the reduced materials requirements and cost to manufacture modules consisting of thin-films as compared to silicon-based wafers.

Other emerging PV technologies include organic, dye-sensitized, quantum-dot, and Perovskite photovoltaics. OPVs fall into the thin-film category of manufacturing, and typically operate around the 12% efficiency range which is lower than the 12–21% typically seen by silicon based PVs. Because organic photovoltaics require very high purity and are relatively reactive they must be encapsulated which vastly increases cost of manufacturing and meaning that they are not feasible for large scale up. Dye-sensitized PVs are similar in efficiency to OPVs but are significantly easier to manufacture. However these dye-sensitized photovoltaics present storage problems because the liquid electrolyte is toxic and can potentially permeate the plastics used in the cell. Quantum dot solar cells are quantum dot sensitized DSSCs and are solution processed meaning they are potentially scalable, but currently they have not reached greater than 10% efficiency. Perovskite solar cells are a very efficient solar energy converter and have excellent optoelectric properties for photovoltaic purposes, but they are expensive and difficult to manufacture.

Applications

Photovoltaic Systems

A photovoltaic system, or solar PV system is a power system designed to supply usable solar power by means of photovoltaics. It consists of an arrangement of several components, including solar panels to absorb and directly convert sunlight into electricity, a solar inverter to change the electric current from DC to AC, as well as mounting, cabling and other electrical accessories. PV systems range from small, roof-top mounted or building-integrated systems with capacities from a few to several tens of kilowatts, to large utility-scale power stations of hundreds of megawatts. Nowadays, most PV systems are grid-connected, while stand-alone systems only account for a small portion of the market.

- Rooftop and building integrated systems

Rooftop PV on half-timbered house

Photovoltaic arrays are often associated with buildings: either integrated into them, mounted on them or mounted nearby on the ground. Rooftop PV systems are most often retrofitted into existing buildings, usually mounted on top of the existing roof structure or on the existing walls. Alternatively, an array can be located separately from the building but connected by cable to supply power for the building. Building-integrated photovoltaics (BIPV) are increasingly incorporated into the roof or walls of new domestic and industrial buildings as a principal or ancillary source of electrical power. Roof tiles with integrated PV cells are sometimes used as well. Provided there is an open gap in which air can circulate, rooftop mounted solar panels can provide a passive cooling effect on buildings during the day and also keep accumulated heat in at night. Typically, residential rooftop systems have small capacities of around 5–10 kW, while commercial rooftop systems often amount to several hundreds of kilowatts. Although rooftop systems are much smaller than ground-mounted utility-scale power plants, they account for most of the worldwide installed capacity.

- Concentrator photovoltaics

Concentrator photovoltaics (CPV) is a photovoltaic technology that contrary to conventional flat-plate PV systems uses lenses and curved mirrors to focus sunlight onto small, but highly efficient, multi-junction (MJ) solar cells. In addition, CPV systems often use solar trackers and sometimes a cooling system to further increase their efficiency. Ongoing research and development is rapidly improving their competitiveness in the utility-scale segment and in areas of high solar insolation.

- Photovoltaic thermal hybrid solar collector

Photovoltaic thermal hybrid solar collector (PVT) are systems that convert solar radiation into thermal and electrical energy. These systems combine a solar PV cell, which converts sunlight into electricity, with a solar thermal collector, which captures the remaining energy and removes waste heat from the PV module. The capture of both electricity and heat allow these devices to have higher exergy and thus be more overall energy efficient than solar PV or solar thermal alone.

- Power stations

Satellite image of the Topaz Solar Farm

Many utility-scale solar farms have been constructed all over the world. As of 2015, the 579-megawatt (MW_{AC}) Solar Star is the world's largest photovoltaic power station, followed by the Desert Sunlight Solar Farm and the Topaz Solar Farm, both with a capacity of 550 MW_{AC}, constructed by US-company First Solar, using CdTe modules, a thin-film PV technology. All three power stations are located in the Californian desert. Many solar farms around the world are integrated with agriculture and some use innovative solar tracking systems that follow the sun's daily path across the sky to generate more electricity than conventional fixed-mounted systems. There are no fuel costs or emissions during operation of the power stations.

- Rural electrification

Developing countries where many villages are often more than five kilometers away from grid power are increasingly using photovoltaics. In remote locations in India a rural lighting program has been providing solar powered LED lighting to replace kerosene lamps. The solar powered lamps were sold at about the cost of a few months' supply of kerosene. Cuba is working to provide solar power for areas that are off grid. More complex applications of off-grid solar energy use include 3D printers. RepRap 3D printers have been solar powered with photovoltaic technology, which enables distributed manufacturing for sustainable development. These are areas where the social costs and benefits offer an excellent case for going solar, though the lack of profitability has relegated such endeavors to humanitarian efforts. However, in 1995 solar rural electrification projects had been found to be difficult to sustain due to unfavorable economics, lack of technical support, and a legacy of ulterior motives of north-to-south technology transfer.

- Standalone systems

Standalone PV system at an ecotourism resort (British Columbia, Canada).

Until a decade or so ago, PV was used frequently to power calculators and novelty devices. Improvements in integrated circuits and low power liquid crystal displays make it possible to power such devices for several years between battery changes, making PV use less common. In contrast, solar powered remote fixed devices have seen increasing use recently in locations where significant connection cost makes grid power prohibitively expensive. Such applications include solar lamps, water pumps, parking meters, emergency telephones, trash compactors, temporary traffic signs, charging stations, and remote guard posts and signals.

- Floatovoltaics

In May 2008, the Far Niente Winery in Oakville, CA pioneered the world's first "floatovoltaic" system by installing 994 photovoltaic solar panels onto 130 pontoons and floating them on the winery's irrigation pond. The floating system generates about 477 kW of peak output and when combined with an array of cells located adjacent to the pond is able to fully offset the winery's electricity consumption. The primary benefit of a floatovoltaic system is that it avoids the need to sacrifice valuable land area that could be used for another purpose. In the case of the Far Niente Winery, the floating system saved three-quarters of an acre that would have been required for a land-based system. That land area can instead be used for agriculture. Another benefit of a floatovoltaic system is that the panels are kept at a lower temperature than they would be on land, leading to a higher efficiency of solar energy conversion. The floating panels also reduce the amount of water lost through evaporation and inhibit the growth of algae.

- In transport

Solar Impulse 2, a solar aircraft

PV has traditionally been used for electric power in space. PV is rarely used to provide motive power in transport applications, but is being used increasingly to provide auxiliary power in boats and cars. Some automobiles are fitted with solar-powered air conditioning to limit interior temperatures on hot days. A self-contained solar vehicle would have limited power and utility, but a solar-charged electric vehicle allows use of solar power for transportation. Solar-powered cars, boats and airplanes have been demonstrated, with the most practical and likely of these being solar cars. The Swiss solar aircraft, Solar Impulse 2, achieved the longest non-stop solo flight in history and plan to make the first solar-powered aerial circumnavigation of the globe in 2015.

- Telecommunication and signaling

Solar PV power is ideally suited for telecommunication applications such as local telephone exchange, radio and TV broadcasting, microwave and other forms of electronic communication links. This is because, in most telecommunication application, storage batteries are already in use and the electrical system is basically DC. In hilly and mountainous terrain, radio and TV signals may not reach as they get blocked or reflected back due to undulating terrain. At these locations, low power transmitters (LPT) are installed to receive and retransmit the signal for local population.

- Spacecraft applications

Part of Juno's solar array

Solar panels on spacecraft are usually the sole source of power to run the sensors, active heating and cooling, and communications. A battery stores this energy for use when the solar panels are in shadow. In some, the power is also used for spacecraft propulsion—electric propulsion. Spacecraft were one of the earliest applications of photovoltaics, starting with the silicon solar cells used on the Vanguard 1 satellite, launched by the US in 1958. Since then, solar power has been used on missions ranging from the MESSENGER probe to Mercury, to as far out in the solar system as the Juno probe to Jupiter. The largest solar power system flown in space is the electrical system of the International Space Station. To increase the power generated per kilogram, typical spacecraft solar panels use high-cost, high-efficiency, and close-packed rectangular multi-junction solar cells made of gallium arsenide (GaAs) and other semiconductor materials.

- Specialty Power Systems

Photovoltaics may also be incorporated as energy conversion devices for objects at elevated temperatures and with preferable radiative emissivities such as heterogeneous combustors.

Advantages

The 122 PW of sunlight reaching the Earth's surface is plentiful—almost 10,000 times more than the 13 TW equivalent of average power consumed in 2005 by humans. This abundance leads to the suggestion that it will not be long before solar energy will become the world's primary energy source. Additionally, solar electric generation has the highest power density (global mean of 170 W/m^2) among renewable energies.

Solar power is pollution-free during use. Production end-wastes and emissions are manageable using existing pollution controls. End-of-use recycling technologies are under development and policies are being produced that encourage recycling from producers.

PV installations can operate for 100 years or even more with little maintenance or intervention after their initial set-up, so after the initial capital cost of building any solar power plant, operating costs are extremely low compared to existing power technologies.

Grid-connected solar electricity can be used locally thus reducing transmission/distribution losses (transmission losses in the US were approximately 7.2% in 1995).

Compared to fossil and nuclear energy sources, very little research money has been invested in the development of solar cells, so there is considerable room for improvement. Nevertheless, experimental high efficiency solar cells already have efficiencies of over 40% in case of concentrating photovoltaic cells and efficiencies are rapidly rising while mass-production costs are rapidly falling.

In some states of the United States, much of the investment in a home-mounted system may be lost if the home-owner moves and the buyer puts less value on the system than the seller. The city of Berkeley developed an innovative financing method to remove this limitation, by adding a tax assessment that is transferred with the home to pay for the solar panels. Now known as PACE, Property Assessed Clean Energy, 30 U.S. states have duplicated this solution.

There is evidence, at least in California, that the presence of a home-mounted solar system can actually increase the value of a home. According to a paper published in April 2011 by the Ernest Orlando Lawrence Berkeley National Laboratory titled An Analysis of the Effects of Residential Photovoltaic Energy Systems on Home Sales Prices in California:

The research finds strong evidence that homes with PV systems in California have sold for a premium over comparable homes without PV systems. More specifically, estimates for average PV premiums range from approximately $3.9 to $6.4 per installed watt (DC) among a large number of different model specifications, with most models coalescing near $5.5/watt. That value corresponds to a premium of approximately $17,000 for a relatively new 3,100 watt PV system (the average size of PV systems in the study).

Photovoltaic System

A photovoltaic system, also solar PV power system, or PV system, is a power system designed to supply usable solar power by means of photovoltaics. It consists of an arrangement of several components, including solar panels to absorb and convert sunlight into electricity, a solar inverter to change the electric current from DC to AC, as well as mounting, cabling and other electrical accessories to set up a working system. It may also use a solar tracking system to improve the system's overall performance and include an integrated battery solution, as prices for storage devices are expected to decline. Strictly speaking, a solar array only encompasses the ensemble of solar panels, the visible part of the PV system, and does not include all the other hardware, often summarized as balance of system (BOS). Moreover, PV systems convert light directly into electricity and shouldn't be confused with other technologies, such as concentrated solar power or solar thermal, used for heating and cooling.

PV systems range from small, rooftop-mounted or building-integrated systems with capacities from a few to several tens of kilowatts, to large utility-scale power stations of hundreds of megawatts. Nowadays, most PV systems are grid-connected, while off-grid or stand-alone systems only account for a small portion of the market.

Operating silently and without any moving parts or environmental emissions, PV systems have developed from being niche market applications into a mature technology used for mainstream electricity generation. A rooftop system recoups the invested energy for its manufacturing and installation within 0.7 to 2 years and produces about 95 percent of net clean renewable energy over a 30-year service lifetime.

Due to the exponential growth of photovoltaics, prices for PV systems have rapidly declined in recent years. However, they vary by market and the size of the system. In 2014, prices for residential 5-kilowatt systems in the United States were around $3.29 per watt, while in the highly penetrated German market, prices for rooftop systems of up to 100 kW declined to €1.24 per watt. Nowadays, solar PV modules account for less than half of the system's overall cost, leaving the rest to the remaining BOS-components and to soft costs, which include customer acquisition, permitting, inspection and interconnection, installation labor and financing costs.

Modern System

Overview

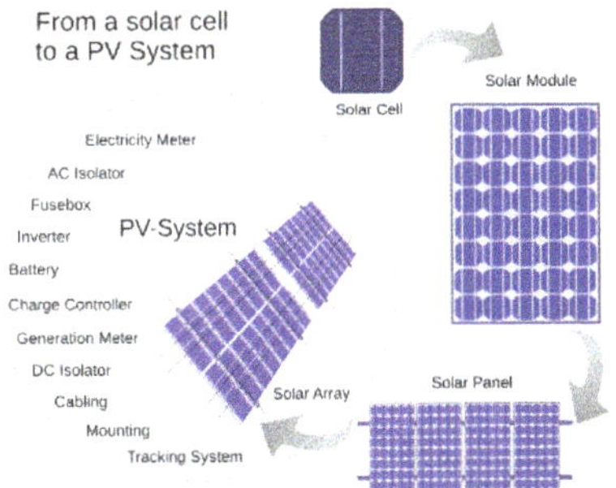

Diagram of the possible components of a photovoltaic system

A photovoltaic system converts the sun's radiation into usable electricity. It comprises the solar array and the balance of system components. PV systems can be categorized by various aspects, such as, grid-connected vs. stand alone systems, building-integrated vs. rack-mounted systems, residential vs. utility systems, distributed vs. centralized systems, rooftop vs. ground-mounted systems, tracking vs. fixed-tilt systems, and new constructed vs. retrofitted systems. Other distinctions may include, systems with microinverters vs. central inverter, systems using crystalline silicon vs. thin-film technology, and systems with modules from Chinese vs. European and U.S.-manufacturers.

About 99 percent of all European and 90 percent of all U.S. solar power systems are connected to the electrical grid, while off-grid systems are somewhat more common in Australia and South Korea. PV systems rarely use battery storage. This may change soon, as government incentives for distributed energy storage are being implemented and investments in storage solutions are gradually becoming economically viable for small systems. A solar array of a typical residential PV system is rack-mounted on the roof, rather than integrated into the roof or facade of the building, as this is significantly more expensive. Utility-scale solar power stations are ground-mounted, with fixed tilted solar panels rather than using expensive tracking devices. Crystalline silicon is the predominant material used in 90 percent of worldwide produced solar modules, while rival thin-film has lost market-share in recent years. About 70 percent of all solar cells and modules are produced in China and Taiwan, leaving only 5 percent to European and US-manufacturers. The installed capacity for both, small rooftop systems and large solar power stations is growing rapidly and in equal parts, although there is a notable trend towards utility-scale systems, as the focus on new installations is shifting away from Europe to sunnier regions, such as the Sunbelt in the U.S., which are less opposed to ground-mounted solar farms and cost-effectiveness is more emphasized by investors.

Driven by advances in technology and increases in manufacturing scale and sophistication, the cost of photovoltaics is declining continuously. There are several million PV systems distributed all over the world, mostly in Europe, with 1.4 million systems in Germany alone as well as North America with 440,000 systems in the United States, The energy conversion efficiency of a conventional solar module increased from 15 to 20 percent over the last 10 years and a PV system recoups the energy needed for its manufacture in about 2 years. In exceptionally irradiated locations, or when thin-film technology is used, the so-called energy payback time decreases to one year or less. Net metering and financial incentives, such as preferential feed-in tariffs for solar-generated electricity, have also greatly supported installations of PV systems in many countries. The levelised cost of electricity from large-scale PV systems has become competitive with conventional electricity sources in an expanding list of geographic regions, and grid parity has been achieved in about 30 different countries.

As of 2015, the fast-growing global PV market is rapidly approaching the 200 GW mark – about 40 times the installed capacity of 2006. Photovoltaic systems currently contribute about 1 percent to worldwide electricity generation. Top installers of PV systems in terms of capacity are currently China, Japan and the United States, while half of the world's capacity is installed in Europe, with Germany and Italy supplying 7% to 8% of their respective domestic electricity consumption with solar PV. The International Energy Agency expects solar power to become the world's largest source of electricity by 2050, with solar photovoltaics and concentrated solar thermal contributing 16% and 11% to the global demand, respectively.

Grid-connection

A grid connected system is connected to a larger independent grid (typically the public electricity grid) and feeds energy directly into the grid. This energy may be shared by a residential or commercial building before or after the revenue measurement point. The difference being whether the credited energy production is calculated independently of the customer's energy consumption (feed-in tariff) or only on the difference of energy (net metering). Grid connected systems vary in size from residential (2–10 kW_p) to solar power stations (up to 10s of MW_p). This is a form of decentralized electricity generation. The feeding of electricity into the grid requires the transformation of DC into AC by a special, synchronising grid-tie inverter. In kilowatt-sized installations the DC side system voltage is as high as permitted (typically 1000V except US residential 600 V) to limit ohmic losses. Most modules (60 or 72 crystalline silicon cells) generate 160 W to 300 W at 36 volts. It is sometimes necessary or desirable to connect the modules partially in parallel rather than all in series. One set of modules connected in series is known as a 'string'.

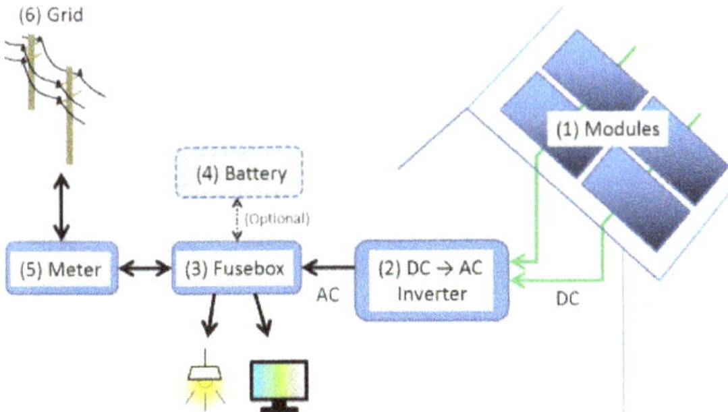

Schematics of a typical residential PV system

Scale of System

Photovoltaic systems are generally categorized into three distinct market segments: residential rooftop, commercial rooftop, and ground-mount utility-scale systems. Their capacities range from a few kilowatts to hundreds of megawatts. A typical residential system is around 10 kilowatts and mounted on a sloped roof, while commercial systems may reach a megawatt-scale and are generally installed on low-slope or even flat roofs. Although rooftop mounted systems are small and display a higher cost per watt than large utility-scale installations, they account for the largest share in the market. There is, however, a growing trend towards bigger utility-scale power plants, especially in the "sunbelt" region of the planet.

Utility-scale

Perovo Solar Park in Ukraine

Large utility-scale solar parks or farms are power stations and capable of providing an energy supply to large numbers of consumers. Generated electricity is fed into the transmission grid powered by central generation plants (grid-connected or grid-tied plant), or combined with one, or many, domestic electricity generators to feed into a small electrical grid (hybrid plant). In rare cases generated electricity is stored or used directly by island/standalone plant. PV systems are generally designed in order to ensure the highest energy yield for a given investment. Some large photovoltaic power stations such as Solar Star, Waldpolenz Solar Park and Topaz Solar Farm cover tens or hundreds of hectares and have power outputs up to hundreds of megawatts.

Rooftop, Mobile, and Portable

Rooftop system near Boston, USA.

A small PV system is capable of providing enough AC electricity to power a single home, or even an isolated device in the form of AC or DC electric. For example, military and civilian Earth observation satellites, street lights, construction and traffic signs, electric cars, solar-powered tents, and electric aircraft may contain integrated photovoltaic systems to provide a primary or auxiliary power source in the form of AC or DC power, depending on the design and power demands. In 2013, rooftop systems accounted for 60 percent of worldwide installations. However, there is a trend away from rooftop and towards utility-scale PV systems, as the focus of new PV installations is also shifting from Europe to countries in the sunbelt region of the planet where opposition to ground-mounted solar farms is less accentuated.

Portable and mobile PV systems provide electrical power independent of utility connections, for "off the grid" operation. Such systems are so commonly used on recreational vehicles and boats that there are retailers specializing in these applications and products specifically targeted to them. Since recreational vehicles (RV) normally carry batteries and operate lighting and other systems on nominally 12-volt DC power, RV PV systems normally operate in a voltage range chosen to charge 12-volt batteries directly, and addition of a PV system requires only panels, a charge controller, and wiring.

Building-integrated

BAPV wall near Barcelona, Spain

In urban and suburban areas, photovoltaic arrays are commonly used on rooftops to supplement power use; often the building will have a connection to the power grid, in which case the energy produced by the PV array can be sold back to the utility in some sort of net metering agreement. Some utilities, such as Solvay Electric in Solvay, NY, use the rooftops of commercial customers and telephone poles to support their use of PV panels. Solar trees are arrays that, as the name implies, mimic the look of trees, provide shade, and at night can function as street lights.

Performance

Uncertainties in revenue over time relate mostly to the evaluation of the solar resource and to the performance of the system itself. In the best of cases, uncertainties are typically 4% for year-to-year climate variability, 5% for solar resource estimation (in a horizontal plane), 3% for estimation of irradiation in the plane of the array, 3% for power rating of modules, 2% for losses due to dirt and soiling, 1.5% for losses due to snow, and 5% for other sources of error. Identifying and reacting to manageable losses is critical for revenue and O&M efficiency. Monitoring of array performance may be part of contractual agreements between the array owner, the builder, and the utility purchasing the energy produced. Recently, a method to create "synthetic days" using readily available weather data and verification using the Open Solar Outdoors Test Field make it possible to predict photovoltaic systems performance with high degrees of accuracy. This method can be used to then determine loss mechanisms on a local scale - such as those from snow or the effects of surface coatings (e.g. hydrophobic or hydrophilic) on soiling or snow losses. (Although in heavy snow environments with severe ground interference can result in annual losses from snow of 30%.) Access to the Internet has allowed a further improvement in energy monitoring and communication. Dedicated systems are available from a number of vendors. For solar PV system that use micro-inverters (panel-level DC to AC conversion), module power data is automatically provided. Some systems allow setting performance alerts that trigger phone/email/text warnings when limits are reached. These solutions provide data for the system owner and the installer. Installers are able to remotely monitor multiple installations, and see at-a-glance the status of their entire installed base.

Components

A photovoltaic system for residential, commercial, or industrial energy supply consists of the solar array and a number of components often summarized as the balance of system (BOS). The term originates from the fact that some BOS-components are balancing the power-generating subsystem of the solar array with the power-using side, the load. BOS-components include power-conditioning equipment and structures for mounting, typically one or more DC to AC power converters, also known as inverters, an energy storage device, a racking system that supports the solar array, electrical wiring and interconnections, and mounting for other components.

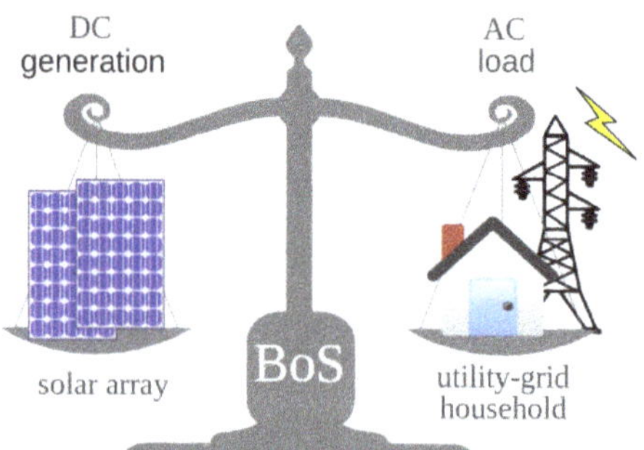

The balance of system components of a PV system (BOS) balance the power-generating subsystem of the solar array (left side) with the power-using side of the AC-household devices and the utility grid (right side).

Optionally, a balance of system may include any or all of the following: renewable energy credit revenue-grade meter, maximum power point tracker (MPPT), battery system and charger, GPS solar tracker, energy management software, solar irradiance sensors, anemometer, or task-specific accessories designed to meet specialized requirements for a system owner. In addition, a CPV system requires optical lenses or mirrors and sometimes a cooling system.

The terms *"solar array"* and *"PV system"* are often used interchangeably, despite the fact that the solar array does not encompass the entire system. Moreover, *"solar panel"* is often used as a synonym for *"solar module"*, although a panel consists of a string of several modules. The term *"solar system"* is also an often used misnomer for a PV system.

Solar Array

Conventional c-Si solar cells, normally wired in series, are encapsulated in a solar module to protect them from the weather. The module consists of a tempered glass as cover, a soft and flexible encapsulant, a rear backsheet made of a weathering and fire-resistant material and an aluminium frame around the outer edge. Electrically connected and mounted on a supporting structure, solar modules build a string of modules, often called solar panel. A solar array consists of one or many such panels. A photovoltaic array, or solar array, is a linked collection of solar panels. The power that one module can produce is seldom enough to meet requirements of a home or a business, so the modules are linked together to form an array. Most PV arrays use an inverter to convert the DC power produced by the modules into alternating current that can power lights, motors, and other loads. The modules in a PV array are usually first connected in series to obtain the desired voltage; the individual strings are then connected in parallel to allow the system to produce more current. Solar panels are typically measured under STC (standard test conditions) or PTC (PVUSA test conditions), in watts. Typical panel ratings range from less than 100 watts to over 400 watts. The array rating consists of a summation of the panel ratings, in watts, kilowatts, or megawatts.

Module and Efficiency

A typical "150 watt" PV module is about a square meter in size. Such a module may be expected to produce 0.75 kilowatt-hour (kWh) every day, on average, after taking into account the weather and the latitude, for an insolation of 5 sun hours/day. In the last 10 years, the efficiency of average commercial wafer-based crystalline silicon modules increased from about 12% to 16% and CdTe module efficiency increased from 9% to 13% during same period. Module output and life degraded by increased temperature. Allowing ambient air to flow over, and if possible behind, PV modules reduces this problem. Effective module lives are typically 25 years or more. The payback period for an investment in a PV solar installation varies greatly and is typically less useful than a calculation of return on investment. While it is typically calculated to be between 10 and 20 years, the financial payback period can be far shorter with incentives.

Fixed tilt solar array in of crystalline silicon panels in Canterbury, New Hampshire, United States

Solar array of a solar farm with a few thousand solar modules on the island of Majorca, Spain

Due to the low voltage of an individual solar cell (typically ca. 0.5V), several cells are wired in series in the manufacture of a "laminate". The laminate is assembled into a protective weatherproof enclosure, thus making a photovoltaic module or solar panel. Modules may then be strung together into a photovoltaic array. In 2012, solar panels available for consumers can have an efficiency of up to about 17%, while commercially available panels can go as far as 27%. It has been recorded that a group from the The Fraunhofer Institute for Solar Energy Systems have created a cell that can reach 44.7% efficiency, which makes scientists' hopes of reaching the 50% efficiency threshold a lot more feasible.

Shading and Dirt

Photovoltaic cell electrical output is extremely sensitive to shading. The effects of this shading are well known. When even a small portion of a cell, module, or array is shaded, while the remainder is in sunlight, the output falls dramatically due to internal 'short-circuiting' (the electrons reversing course through the shaded portion of the p-n junction). If the current drawn from the series string of cells is no greater than the current that can be produced by the shaded cell, the current (and so power) developed by the string is limited. If enough voltage is available from the rest of the cells in a string, current will be forced through the cell by breaking down the junction in the shaded portion. This breakdown voltage in common cells is between 10 and 30 volts. Instead of adding to the power produced by the panel, the shaded cell absorbs power, turning it into heat. Since the reverse voltage of a shaded cell is much greater than the forward voltage of an illuminated cell, one shaded cell can absorb the power of many other cells in the string, disproportionately affecting panel output. For example, a shaded cell may drop 8 volts, instead of adding 0.5 volts, at a particular current level, thereby absorbing the power produced by 16 other cells. It is, thus important that a PV installation is not shaded by trees or other obstructions.

Several methods have been developed to determine shading losses from trees to PV systems over both large regions using LiDAR, but also at an individual system level using sketchup. Most modules have bypass diodes between each cell or string of cells that minimize the effects of shading and only lose the power of the shaded portion of the array. The main job of the bypass diode is to eliminate hot spots that form on cells that can cause further damage to the array, and cause fires. Sunlight can be absorbed by dust, snow, or other impurities at the surface of the module. This can reduce the light that strikes the cells. In general these losses aggregated over the year are small even for locations in Canada. Maintaining a clean module surface will increase output performance over the life of the module. Google found that cleaning the flat mounted solar panels after 15 months increased their output by almost 100%, but that the 5% tilted arrays were adequately cleaned by rainwater.

Insolation and Energy

Solar insolation is made up of direct, diffuse, and reflected radiation. The absorption factor of a PV cell is defined as the fraction of incident solar irradiance that is absorbed by the cell. At high noon on a cloudless day at the equator, the power of the sun is about 1 kW/m^2, on the Earth's surface, to a plane that is perpendicular to the sun's rays. As such, PV arrays can track the sun through each day to greatly enhance energy collection. However, tracking devices add cost, and require maintenance, so it is more common for PV arrays to have fixed mounts that tilt the array and face solar noon (approximately due south in the Northern Hemisphere or due north in the Southern Hemisphere). The tilt angle, from horizontal, can be varied for season, but if fixed, should be set to give optimal array output during the peak electrical demand portion of a typical year for a stand-alone system. This optimal module tilt angle is not necessarily identical to the tilt angle for maximum annual array energy output. The optimization of the a photovoltaic system for a specific environment can be complicated as issues of solar flux, soiling, and snow losses should be taken into effect. In addition, recent work has shown that spectral effects can play a role in optimal photovoltaic material selection. For example, the spectral albedo can play a significant role in output depending on the surface around the photovoltaic system and the type of solar cell material. For the weather and latitudes of the United States and Europe, typical insolation ranges from 4 kWh/m^2/day in northern climes to 6.5 kWh/m^2/day in the sunniest regions. A photovoltaic installation in the southern latitudes of Europe or the United States may expect to produce 1 kWh/m^2/day. A typical 1 kW photovoltaic installation in Australia or the southern latitudes of Europe or United States, may produce 3.5–5 kWh per day, dependent on location, orientation, tilt, insolation and other factors. In the Sahara desert, with less cloud cover and a better solar angle, one could ideally obtain closer to 8.3 kWh/m^2/day provided the nearly ever present wind would not blow sand onto the units. The area of the Sahara desert is over 9 million km^2. 90,600 km^2, or about 1%, could generate as much electricity as all of the world's power plants combined.

Mounting

Modules are assembled into arrays on some kind of mounting system, which may be classified as ground mount, roof mount or pole mount. For solar parks a large rack is mounted on the ground, and the modules mounted on the rack. For buildings, many different racks have been devised for pitched roofs. For flat roofs, racks, bins and building integrated solutions are used. Solar panel racks mounted on top of poles can be stationary or moving. Side-of-pole mounts are suitable for situations where a pole has something else mounted at its top, such as a light fixture or an antenna. Pole mounting raises what would otherwise be a ground mounted array above weed shadows and livestock, and may satisfy electrical code requirements regarding inaccessibility of exposed wiring. Pole mounted panels are open to more cooling air on their underside, which increases performance. A multiplicity of pole top racks can be formed into a parking carport or other shade structure. A rack which does not follow the sun from left to right may allow seasonal adjustment up or down.

A 23-year-old, ground mounted PV system from the 1980s on a North Frisian Island, Germany. The modules conversion efficiency was only 12%.

Cabling

Due to their outdoor usage, solar cables are specifically designed to be resistant against UV radiation and extremely high temperature fluctuations and are generally unaffected by the weather. A number of standards specify the usage of electrical wiring in PV systems, such as the IEC 60364 by the International Electrotechnical Commission, in section 712 *"Solar photovoltaic (PV) power supply systems"*, the British Standard BS 7671, incorporating regulations relating to microgeneration and photovoltaic systems, and the US UL4703 standard, in subject 4703 *"Photovoltaic Wire"*.

Tracker

A solar tracking system tilts a solar panel throughout the day. Depending on the type of tracking system, the panel is either aimed directly at the sun or the brightest area of a partly clouded sky. Trackers greatly enhance early morning and late afternoon performance, increasing the total amount of power produced by a system by about 20–25% for a single axis tracker and about 30% or more for a dual axis tracker, depending on latitude. Trackers are effective in regions that receive a large portion of sunlight directly. In diffuse light (i.e. under cloud or fog), tracking has little or no value. Because most concentrated photovoltaics systems are very sensitive to the sunlight's angle, tracking systems allow them to produce useful power for more than a brief period each day. Tracking systems improve performance for two main reasons. First, when a solar panel is perpendicular to the sunlight, it receives more light on its surface than if it were angled. Second, direct light is used more efficiently than angled light. Special Anti-reflective coatings can improve solar panel efficiency for direct and angled light, somewhat reducing the benefit of tracking.

A 1998 model of a passive solar tracker, viewed from underneath.

Trackers and sensors to optimise the performance are often seen as optional, but tracking systems can increase viable output by up to 45%. PV arrays that approach or exceed one megawatt often use solar trackers. Accounting for clouds, and the fact that most of the world is not on the equator, and that the sun sets in the evening, the correct measure of solar power is insolation – the average number of kilowatt-hours per square meter per day. For the weather and latitudes of the United States and Europe, typical insolation ranges from 2.26 kWh/m^2/day in northern climes to 5.61 kWh/m^2/day in the sunniest regions.

For large systems, the energy gained by using tracking systems can outweigh the added complexity (trackers can increase efficiency by 30% or more). For very large systems, the added maintenance of tracking is a substantial detriment. Tracking is not required for flat panel and low-concentration photovoltaic systems. For high-concentration photovoltaic systems, dual axis tracking is a necessity. Pricing trends affect the balance between adding more stationary solar panels versus having fewer panels that track. When solar panel prices drop, trackers become a less attractive option.

Inverter

Systems designed to deliver alternating current (AC), such as grid-connected applications need an inverter to convert the direct current (DC) from the solar modules to AC. Grid connected inverters must supply AC electricity in sinusoidal form, synchronized to the grid frequency, limit feed in voltage to no higher than the grid voltage and disconnect from the grid if the grid voltage is turned off. Islanding inverters need only produce regulated voltages and frequencies in a sinusoidal waveshape as no synchronisation or co-ordination with grid supplies is required.

Central inverter with AC and DC disconnects (on the side), monitoring gateway, transformer isolation and interactive LCD.

String inverter (left), generation meter, and AC disconnect (right). A modern 2013 installation in Vermont, United States.

A solar inverter may connect to a string of solar panels. In some installations a solar micro-inverter is connected at each solar panel. For safety reasons a circuit breaker is provided both on the AC and DC side to enable maintenance. AC output may be connected through an electricity meter into the public grid. The number of modules in the system determines the total DC watts capable of being generated by the solar array; however, the inverter ultimately governs the amount of AC watts that can be distributed for consumption. For example, a PV system comprising 11 kilowatts DC (kW_{DC})

worth of PV modules, paired with one 10-kilowatt AC (kW_{AC}) inverter, will be limited to the inverter's output of 10 kW. As of 2014, conversion efficiency for state-of-the-art converters reached more than 98 percent. While string inverters are used in residential to medium-sized commercial PV systems, central inverters cover the large commercial and utility-scale market. Market-share for central and string inverters are about 50 percent and 48 percent, respectively, leaving less than 2 percent to micro-inverters.

Maximum power point tracking (MPPT) is a technique that grid connected inverters use to get the maximum possible power from the photovoltaic array. In order to do so, the inverter's MPPT system digitally samples the solar array's ever changing power output and applies the proper resistance to find the optimal *maximum power point*.

Anti-islanding is a protection mechanism that immediately shuts down the inverter preventing it from generating AC power when the connection to the load no longer exists. This happens, for example, in the case of a blackout. Without this protection, the supply line would become an "island" with power surrounded by a "sea" of unpowered lines, as the solar array continues to deliver DC power during the power outage. Islanding is a hazard to utility workers, who may not realize that an AC circuit is still powered, and it may prevent automatic re-connection of devices.

Battery

Although still expensive, PV systems increasingly use rechargeable batteries to store a surplus to be later used at night. Batteries used for grid-storage also stabilize the electrical grid by leveling out peak loads, and play an important role in a smart grid, as they can charge during periods of low demand and feed their stored energy into the grid when demand is high.

Common battery technologies used in today's PV systems include the valve regulated lead-acid battery– a modified version of the conventional lead–acid battery, nickel–cadmium and lithium-ion batteries. Compared to the other types, lead-acid batteries have a shorter lifetime and lower energy density. However, due to their high reliability, low self discharge as well as low investment and maintenance costs, they are currently the predominant technology used in small-scale, residential PV systems, as lithium-ion batteries are still being developed and about 3.5 times as expensive as lead-acid batteries. Furthermore, as storage devices for PV systems are stationary, the lower energy and power density and therefore higher weight of lead-acid batteries are not as critical as, for example, in electric transportation Other rechargeable batteries that are considered for distributed PV systems include sodium–sulfur and vanadium redox batteries, two prominent types of a molten salt and a flow battery, respectively. In 2015, Tesla motors launched the Powerwall, a rechargeable lithium-ion battery with the aim to revolutionize energy consumption.

PV systems with an integrated battery solution also need a charge controller, as the varying voltage and current from the solar array requires constant adjustment to prevent damage from overcharging. Basic charge controllers may simply turn the PV panels on and off, or may meter out pulses of energy as needed, a strategy called PWM or pulse-width modulation. More advanced charge controllers will incorporate MPPT logic into their battery charging algorithms. Charge controllers may also divert energy to some purpose other than battery charging. Rather than simply shut off the free PV energy when not needed, a user may choose to heat air or water once the battery is full.

Monitoring and Metering

The metering must be able to accumulate energy units in both directions or two meters must be used. Many meters accumulate bidirectionally, some systems use two meters, but a unidirectional meter (with detent) will not accumulate energy from any resultant feed into the grid. In some countries, for installations over 30 kW_p a frequency and a voltage monitor with disconnection of all phases is required. This is done where more solar power is being generated than can be accommodated by the utility, and the excess can not either be exported or stored. Grid operators historically have needed to provide transmission lines and generation capacity. Now they need to also provide storage. This is normally hydro-storage, but other means of storage are used. Initially storage was used so that baseload generators could operate at full output. With variable renewable energy, storage is needed to allow power generation whenever it is available, and consumption whenever it is needed.

A Canadian electricity meter

The two variables a grid operator have are storing electricity for *when* it is needed, or transmitting it to *where* it is needed. If both of those fail, installations over 30kWp can automatically shut down, although in practice all inverters maintain voltage regulation and stop supplying power if the load is inadequate. Grid operators have the option of curtailing excess generation from large systems, although this is more commonly done with wind power than solar power, and results in a substantial loss of revenue. Three-phase inverters have the unique option of supplying reactive power which can be advantageous in matching load requirements.

Photovoltaic systems need to be monitored to detect breakdown and optimize their operation. There are several *photovoltaic monitoring* strategies depending on the output of the installation and its nature. Monitoring can be performed on site or remotely. It can measure production only, retrieve all the data from the inverter or retrieve all of the data from the communicating equipment (probes, meters, etc.). Monitoring tools can be dedicated to supervision only or offer additional functions. Individual inverters and battery charge controllers may include monitoring using manufacturer specific protocols and software. Energy metering of an inverter may be of limited accuracy and not suitable for revenue metering purposes. A third-party data acquisition system can monitor multiple inverters, using the inverter manufacturer's protocols, and also acquire weather-related information. Independent smart meters may measure the total energy production of a PV array system. Separate measures such as satellite image analysis or a solar radiation meter (a pyranometer) can be used to estimate total insolation for comparison. Data collected from a monitoring system can be displayed remotely over the World Wide Web, such as OSOTF.

Other Systems

This section includes systems that are either highly specialized and uncommon or still an emerging new technology with limited significance. However, standalone or off-grid systems take a special place. They were the most common type of systems during the 1980s and 1990s, when PV technology was still very expensive and a pure niche market of small scale applications. Only in places where no electrical grid was available, they were economically viable. Although new stand-alone systems are still being deployed all around the world, their contribution to the overall installed photovoltaic capacity is decreasing. In Europe, off-grid systems account for 1 percent of installed capacity. In the United States, they account for about 10 percent. Off-grid systems are still common in Australia and South Korea, and in many developing countries.

CPV

Concentrator photovoltaic (CPV) in Catalonia, Spain

Concentrator photovoltaics (CPV) and *high concentrator photovoltaic* (HCPV) systems use optical lenses or curved mirrors to concentrate sunlight onto small but highly efficient solar cells. Besides concentrating optics, CPV systems sometime use solar trackers and cooling systems and are more expensive.

Especially HCPV systems are best suited in location with high solar irradiance, concentrating sunlight up to 400 times or more, with efficiencies of 24–28 percent, exceeding those of regular systems. Various designs of CPV and HCPV systems are commercially available but not very common. However, ongoing research and development is taking place.

CPV is often confused with CSP (concentrated solar power) that does not use photovoltaics. Both technologies favor locations that receive much sunlight and are directly competing with each other.

Hybrid

A hybrid system combines PV with other forms of generation, usually a diesel generator. Biogas is also used. The other form of generation may be a type able to modulate power output as a function of demand. However more than one renewable form of energy may be used e.g. wind. The photovoltaic power generation serves to reduce the consumption of non renewable fuel. Hybrid systems are most often found on islands. Pellworm island in Germany and Kythnos island in Greece are notable examples (both are combined with wind). The Kythnos plant has reduced diesel consumption by 11.2%.

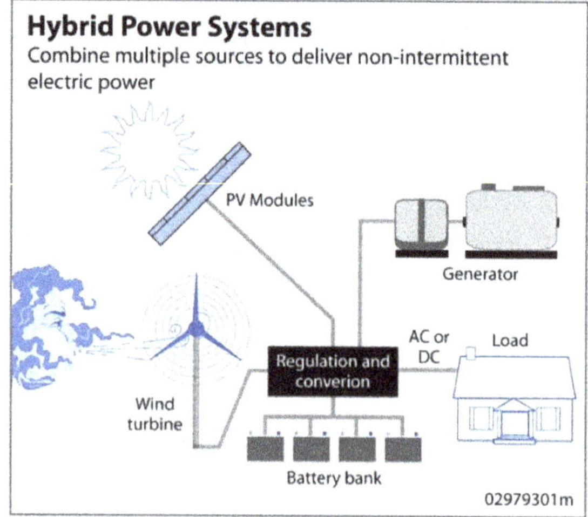

A wind-solar PV hybrid system

In 2015, a case-study conducted in seven countries concluded that in all cases generating costs can be reduced by hybridising mini-grids and isolated grids. However, financing costs for such hybrids are crucial and largely depend on the ownership structure of the power plant. While cost reductions for state-owned utilities can be significant, the study also identified economic benefits to be insignificant or even negative for non-public utilities, such as independent power producers.

There has also been recent work showing that the PV penetration limit can be increased by deploying a distributed network of PV+CHP hybrid systems in the U.S. The temporal distribution of solar flux, electrical and heating requirements for representative

U.S. single family residences were analyzed and the results clearly show that hybridizing CHP with PV can enable additional PV deployment above what is possible with a conventional centralized electric generation system. This theory was reconfirmed with numerical simulations using per second solar flux data to determine that the necessary battery backup to provide for such a hybrid system is possible with relatively small and inexpensive battery systems. In addition, large PV+CHP systems are possible for institutional buildings, which again provide back up for intermittent PV and reduce CHP runtime.

- PVT system (hybrid PV/T), also known as *photovoltaic thermal hybrid solar collectors* convert solar radiation into thermal and electrical energy. Such a system combines a solar (PV) module with a solar thermal collector in an complementary way.
- CPVT system. A *concentrated photovoltaic thermal hybrid* (CPVT) system is similar to a PVT system. It uses concentrated photovoltaics (CPV) instead of conventional PV technology, and combines it with a solar thermal collector.
- CPV/CSP system. A novel solar CPV/CSP hybrid system has been proposed recently, combining concentrator photovoltaics with the non-PV technology of concentrated solar power (CSP), or also known as concentrated solar thermal.
- PV diesel system. It combines a photovoltaic system with a diesel generator. Combinations with other renewables are possible and include wind turbines.

Floating Solar Arrays

Floating solar arrays are PV systems that float on the surface of drinking water reservoirs, quarry lakes, irrigation canals or remediation and tailing ponds. A small number of such systems exist in France, India, Japan, South Korea, the United Kingdom, Singapore and the United States.

The systems are said to have advantages over photovoltaics on land. The cost of land is more expensive, and there are fewer rules and regulations for structures built on bodies of water not used for recreation. Unlike most land-based solar plants, floating arrays can be unobtrusive because they are hidden from public view. They achieve higher efficiencies than PV panels on land, because water cools the panels. The panels have a special coating to prevent rust or corrosion.

In May 2008, the Far Niente Winery in Oakville, California, pioneered the world's first floatovoltaic system by installing 994 solar PV modules with a total capacity of 477 kW onto 130 pontoons and floating them on the winery's irrigation pond. The primary benefit of such a system is that it avoids the need to sacrifice valuable land area that could be used for another purpose. In the case of the Far Niente Winery, it saved three-quarters of an acre that would have been required for a land-based system. Another benefit of a floatovoltaic system is that the panels are kept at a cooler temperature than they would be on land, leading to a higher efficiency of solar energy conversion. The floating PV array also reduces the amount of water lost through evaporation and inhibits the growth of algae.

Utility-scale floating PV farms are starting to be built. The multinational electronics and ceramics manufacturer Kyocera will develop the world's largest, a 13.4 MW farm on the reservoir above Yamakura Dam in Chiba Prefecture using 50,000 solar panels. Salt-water resistant floating farms are also being considered for ocean use, with experiments in Thailand. The largest so far announced floatovoltaic project is a 350 MW power station in the Amazon region of Brazil.

Direct Current Grid

DC grids are found in electric powered transport: railways trams and trolleybuses. A few pilot plants for such applications have been built, such as the tram depots in Hannover Leinhausen, using photovoltaic contributors and Geneva (Bachet de Pesay).The 150 kW_p Geneva site feeds 600V DC directly into the tram/trolleybus electricity network whereas before it provided about 15% of the electricity at its opening in 1999.

Standalone

A stand-alone or off-grid system is not connected to the electrical grid. Standalone systems vary widely in size and application from wristwatches or calculators to remote buildings or spacecraft. If the load is to be supplied independently of solar insolation, the generated power is stored and buffered with a battery. In non-portable applications where weight is not an issue, such as in buildings, lead acid batteries are most commonly used for their low cost and tolerance for abuse.

An isolated mountain hut in Catalonia, Spain

A charge controller may be incorporated in the system to avoid battery damage by excessive charging or discharging. It may also help to optimize production from the solar array using a maximum power point tracking technique (MPPT). However, in simple PV systems where the PV module voltage is matched to the battery voltage, the use of

MPPT electronics is generally considered unnecessary, since the battery voltage is stable enough to provide near-maximum power collection from the PV module. In small devices (e.g. calculators, parking meters) only direct current (DC) is consumed. In larger systems (e.g. buildings, remote water pumps) AC is usually required. To convert the DC from the modules or batteries into AC, an inverter is used.

Solar parking meter in Edinburgh, Scotland

In agricultural settings, the array may be used to directly power DC pumps, without the need for an inverter. In remote settings such as mountainous areas, islands, or other places where a power grid is unavailable, solar arrays can be used as the sole source of electricity, usually by charging a storage battery. Stand-alone systems closely relate to microgeneration and distributed generation.

- Pico PV systems

  The smallest, often portable photovoltaic systems are called pico solar PV systems, or pico solar. They mostly combine a rechargeable battery and charge controller, with a very small PV panel. The panel's nominal capacity is just a few watt-peak (1–10 W_p) and its area less than a tenth of a square meter, or one square foot, in size. A large range of different applications can be solar powered such as music players, fans, portable lamps, security lights, solar lighting kits, solar lanterns and street light, phone chargers, radios, or even small, seven-inch LCD televisions, that run on less than ten watts. As it is the case for power generation from pico hydro, pico PV systems are useful in small, rural communities that require only a small amount of electricity. Since the efficiency of many appliances have improved considerably, in particular due to the usage of LED lights and efficient rechargeable batteries, pico solar has become an affordable alternative, especially in the developing world. The metric prefix *pico-* stands for a *trillionth* to indicate the smallness of the system's electric power.

- Solar street lights

  Solar street lights raised light sources which are powered by photovoltaic panels generally mounted on the lighting structure. The solar array of such off-grid PV system charges a rechargeable battery, which powers a fluorescent or LED lamp during the night. Solar street lights are stand-alone power systems, and have the advantage of savings on trenching, landscaping, and maintenance costs, as well as on the electric bills, despite their higher initial cost compared to conventional street lighting. They are designed with sufficiently large batteries to ensure operation for at least a week and even in the worst situation, they are expected to dim only slightly.

- Telecommunication and signaling

  Solar PV power is ideally suited for telecommunication applications such as local telephone exchange, radio and TV broadcasting, microwave and other forms of electronic communication links. This is because, in most telecommunication application, storage batteries are already in use and the electrical system is basically DC. In hilly and mountainous terrain, radio and TV signals may not reach as they get blocked or reflected back due to undulating terrain. At these locations, low power transmitters are installed to receive and retransmit the signal for local population.

- Solar vehicles

  Solar vehicle, whether ground, water, air or space vehicles may obtain some or all of the energy required for their operation from the sun. Surface vehicles generally require higher power levels than can be sustained by a practically sized solar array, so a battery assists in meeting peak power demand, and the solar array recharges it. Space vehicles have successfully used solar photovoltaic systems for years of operation, eliminating the weight of fuel or primary batteries.

- Solar pumps

  One of the most cost effective solar applications is a solar powered pump, as it is far cheaper to purchase a solar panel than it is to run power lines. They often meet a need for water beyond the reach of power lines, taking the place of a windmill or windpump. One common application is the filling of livestock watering tanks, so that grazing cattle may drink. Another is the refilling of drinking water storage tanks on remote or self-sufficient homes.

- Spacecraft

  Solar panels on spacecraft have been one of the first applications of photovoltaics since the launch of Vanguard 1 in 1958, the first satellite to use solar cells. Contrary to Sputnik, the first artificial satellite to orbit the planet, that ran out of batteries within 21 days due to the lack of solar-power, most modern communications satellites and space probes in the inner solar system rely on the use of solar panels to derive electricity from sunlight.

- Do it yourself community

With a growing interest in environmentally friendly green energy, an increasing number of hobbyists in the DIY-community have endeavored to build their own solar PV systems from kits or partly DIY. Usually, the DIY-community uses inexpensive or high efficiency systems (such as those with solar tracking) to generate their own power. As a result, the DIY-systems often end up cheaper than their commercial counterparts. Often, the system is also hooked up into the regular power grid, using net metering instead of a battery for backup. These systems usually generate power amount of ~2 kW or less. Through the internet, the community is now able to obtain plans to (partly) construct the system and there is a growing trend toward building them for domestic requirements.

Gallery of Standalone Systems

Profile picture of a mobile solar powered generator

Solar panels on a small yacht to charge 12 volt batteries up to 9 amps

Costs and Economy

Median installed system prices for residential PV Systems in Japan, Germany and the United States ($/W)

The cost of producing photovotaic cells have dropped due to economies of scale in production and technological advances in manufacturing. For large-scale installations, prices below $1.00 per watt were common by 2012. A price decrease of 50% had been achieved in Europe from 2006 to 2011 and there is a potential to lower the generation cost by 50% by 2020. Crystal silicon solar cells have largely been replaced by less expensive multicrystalline silicon solar cells, and thin film silicon solar cells have also been developed recently at lower costs of production. Although they are reduced in energy conversion efficiency from single crystalline "siwafers", they are also much easier to produce at comparably lower costs.

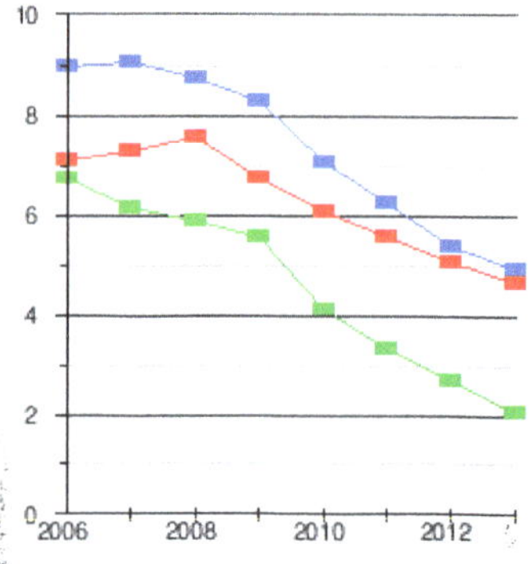

History of solar rooftop prices since 2006. Comparison in US$ per installed watt.

The table below shows the total cost in US cents per kWh of electricity generated by a photovoltaic system. The row headings on the left show the total cost, per peak kilowatt (kW_p), of a photovoltaic installation. Photovoltaic system costs have been declining and in Germany, for example, were reported to have fallen to USD 1389/kW_p by the end of 2014. The column headings across the top refer to the annual energy output in kWh expected from each installed kW_p. This varies by geographic region because the average insolation depends on the average cloudiness and the thickness of atmosphere traversed by the sunlight. It also depends on the path of the sun relative to the panel and the horizon. Panels are usually mounted at an angle based on latitude, and often they are adjusted seasonally to meet the changing solar declination. Solar tracking can also be utilized to access even more perpendicular sunlight, thereby raising the total energy output.

The calculated values in the table reflect the total cost in cents per kWh produced. They assume a 10% total capital cost (for instance 4% interest rate, 1% operating and maintenance cost, and depreciation of the capital outlay over 20 years). Normally, photovoltaic modules have a 25-year warranty.

System Cost 2013

In its 2014 edition of the "Technology Roadmap: Solar Photovoltaic Energy" report, the International Energy Agency (IEA) published prices in US$ per watt for residential, commercial and utility-scale PV systems for eight major markets in 2013.

Typical PV system prices in 2013 in selected countries (USD)								
USD/W	**Austra-lia**	**China**	**France**	**Germany**	**Italy**	**Japan**	**United Kingdom**	**United States**
Residential	1.8	1.5	4.1	2.4	2.8	4.2	2.8	4.9
Commercial	1.7	1.4	2.7	1.8	1.9	3.6	2.4	4.5
Utility-scale	2.0	1.4	2.2	1.4	1.5	2.9	1.9	3.3
Source: IEA – Technology Roadmap: Solar Photovoltaic Energy report								

Regulation

Standardization

Increasing use of photovoltaic systems and integration of photovoltaic power into existing structures and techniques of supply and distribution increases the value of general standards and definitions for photovoltaic components and systems. The standards are compiled at the International Electrotechnical Commission (IEC) and apply to efficiency, durability and safety of cells, modules, simulation programs, plug connectors and cables, mounting systems, overall efficiency of inverters etc.

Planning and Permit

While article 690 of the National Electric Code provides general guidelines for the installation of photovoltaic systems, these guidelines may be superseded by local laws and regulations. Often a permit is required necessitating plan submissions and structural calculations before work may begin. Additionally, many locales require the work to be performed under the guidance of a licensed electrician. Check with the local City/County AHJ (Authority Having Jurisdiction) to ensure compliance with any applicable laws or regulations.

In the United States, the Authority Having Jurisdiction (AHJ) will review designs and issue permits, before construction can lawfully begin. Electrical installation practices must comply with standards set forth within the National Electrical Code (NEC) and be inspected by the AHJ to ensure compliance with building code, electrical code, and fire safety code. Jurisdictions may require that equipment has been tested, certified, listed, and labeled by at least one of the Nationally Recognized Testing Laboratories (NRTL). Despite the complicated installation process, a recent list of solar contractors shows a majority of installation companies were founded since 2000.

National Regulations

United Kingdom

In the UK, PV installations are generally considered permitted development and don't require planning permission. If the property is listed or in a designated area (National Park, Area of Outstanding Natural Beauty, Site of Special Scientific Interest or Norfolk Broads) then planning permission is required.

United States

In the US, many localities require a permit to install a photovoltaic system. A grid-tied system normally requires a licensed electrician to make the connection between the system and the grid-connected wiring of the building. Installers who meet these qualifications are located in almost every state. The State of California prohibits homeowners' associations from restricting solar devices.

Spain

Although Spain generates around 40% of its electricity via photovoltaic and other renewable energy sources, and cities such as Huelva and Seville boast nearly 3,000 hours of sunshine per year, Spain has issued a solar tax to account for the debt created by the investment done by the Spanish government. Those who do not connect to the grid can face up to a fine of 30 million euros ($40 million USD).

Photovoltaic Power Station

A photovoltaic power station, also known as a solar park, is a large-scale photovoltaic system (PV system) designed for the supply of merchant power into the electricity grid. They are differentiated from most building-mounted and other decentralised solar power applications because they supply power at the utility level, rather than to a local user or users. They are sometimes also referred to as solar farms or solar ranches, especially when sited in agricultural areas. The generic expression utility-scale solar is sometimes used to describe this type of project.

The solar power source is via photovoltaic modules that convert light directly to electricity. However, this differs from, and should not be confused with concentrated solar power, the other large-scale solar generation technology, which uses heat to drive a variety of conventional generator systems. Both approaches have their own advantages and disadvantages, but to date, for a variety of reasons, photovoltaic technology has seen much wider use in the field. As of 2013, PV systems outnumber concentrators by about 40 to 1.

The 25.7 MW Lauingen Energy Park in Bavarian Swabia, Germany

In some countries, the nameplate capacity of a photovoltaic power stations is rated in megawatt-peak (MW_p), which refers to the solar array's DC power output. However, Canada, Japan, Spain and some parts of the United States often specify using the converted lower nominal power output in MW_{AC}; a measure directly comparable to other forms of power generation. A third and less common rating is the mega volt-amperes (MVA). Most solar parks are developed at a scale of at least 1 MW_p. As of 2015, the world's largest operating photovoltaic power stations have capacities of close to 600 megawatts and projects up to 1 gigawatt are planned. As at the end of 2015, about 3,400 projects with a combined capacity of 60 GW_{AC} were solar farms larger than 4 MW.

Most of the existing large-scale photovoltaic power stations are owned and operated by independent power producers, but the involvement of community- and utility-owned projects is increasing. To date, almost all have been supported at least in part by regulatory incentives such as feed-in tariffs or tax credits, but as levelized costs have fallen significantly in the last decade and grid parity has been reached in an increasing number of markets, it may not be long before external incentives cease to exist.

History

The first 1 MW_p solar park was built by Arco Solar at Lugo near Hesperia, California at the end of 1982, followed in 1984 by a 5.2 MW_p installation in Carrizo Plain. Both have since been decommissioned, though Carrizo Plain is the site for several large plants now being constructed or planned. The next stage followed the 2004 revisions to the feed-in tariffs in Germany when a substantial volume of solar parks were constructed.

Several hundred installations over 1 MW_p have been since been installed in Germany, of which more than 50 are over 10 MW_p. With its introduction of feed-in tariffs in 2008, Spain became briefly the largest market, with some 60 solar parks over 10 MW, but these incentives have since been withdrawn. The USA, China India, France, Canada, and Italy, amongst others, have also become major markets as shown on the list of photovoltaic power stations.

Serpa Solar Park built in Portugal in 2006

The largest sites under construction have capacities of hundreds of MW_p and projects at a scale of 1 GW_p are being planned.

Siting and Land Use

The land area required for a desired power output, varies depending on the location, and on the efficiency of the solar modules, the slope of the site and the type of mounting used. Fixed tilt solar arrays using typical modules of about 15% efficiency on horizontal sites, need about 1 hectare/MW in the tropics and this figure rises to over 2 hectares in northern Europe.

Because of the longer shadow the array casts when tilted at a steeper angle, this area is typically about 10% higher for an adjustable tilt array or a single axis tracker, and 20% higher for a 2-axis tracker, though these figures will vary depending on the latitude and topography.

The best locations for solar parks in terms of land use are held to be brown field sites, or where there is no other valuable land use. Even in cultivated areas, a significant proportion of the site of a solar farm can also be devoted to other productive uses, such as crop growing or biodiversity.

Agrivoltaics

Agrivoltaics is co-developing the same area of land for both solar photovoltaic power as well as for conventional agriculture. A recent study found that the value of solar generated electricity coupled to shade-tolerant crop production created an over 30% increase in economic value from farms deploying agrivoltaic systems instead of conventional agriculture.

Co-location

In some cases several different solar power stations, with separate owners and contractors, are developed on adjacent sites. This can offer the advantage of the projects sharing the cost and risks of project infrastructure such as grid connections and planning approval. Solar farms can also be co-located with wind farms. Sometimes the title 'solar park' is used, rather than an individual solar power station.

Some examples of such solar clusters are the Charanka Solar Park, where there are 17 different generation projects; Neuhardenberg, with eleven plants, and the Golmud solar parks with total reported capacity over 500MW. An extreme example is calling all of the solar farms in the Gujarat state of India a single solar park, the Gujarat Solar Park.

Technology

Most Solar parks are ground mounted PV systems, also known as free-field solar power plants. They can either be fixed tilt or use a single axis or dual axis solar tracker. While tracking improves the overall performance, it also increases the system's installation and maintenance cost. A solar inverter converts the array's power output from DC to AC, and connection to the utility grid is made through a high voltage, three phase step up transformer of typically 10 kV and above.

Solar Array Arrangements

The solar arrays are the subsystems which convert incoming light into electrical energy. They comprise a multitude of solar modules, mounted on support structures and interconnected to deliver a power output to electronic power conditioning subsystems.

A minority of utility-scale solar parks are configured on buildings and so use building-mounted solar arrays. The majority are 'free field' systems using ground-mounted structures, usually of one of the following types:

Fixed Arrays

Many projects use mounting structures where the solar modules are mounted at a fixed inclination calculated to provide the optimum annual output profile. The modules are normally oriented towards the Equator, at a tilt angle slightly less than the latitude of the site. In some cases, depending on local climatic, topographical or electricity pricing regimes, different tilt angles can be used, or the arrays might be offset from the normal East-West axis to favour morning or evening output.

A variant on this design is the use of arrays, whose tilt angle can be adjusted twice or four times annually to optimise seasonal output. They also require more land area to reduce internal shading at the steeper winter tilt angle. Because the increased output is typically only a few percent, it seldom justifies the increased cost and complexity of this design.

Dual Axis Trackers

To maximise the intensity of incoming direct radiation, solar panels should be orientated normal to the sun's rays. To achieve this, arrays can be designed using two-axis trackers, capable of tracking the sun in its daily orbit across the sky, and as its elevation changes throughout the year.

Bellpuig Solar Park near Lerida, Spain uses pole-mounted 2-axis trackers

These arrays need to be spaced out to reduce inter-shading as the sun moves and the array orientations change, so need more land area. They also require more complex mechanisms to maintain the array surface at the required angle. The increased output can be of the order of 30% in locations with high levels of direct radiation, but the increase is lower in temperate climates or those with more significant diffuse radiation, due to overcast conditions. For this reason, dual axis trackers are most commonly used in subtropical regions, and were first deployed at utility scale at the Lugo plant.

Single Axis Trackers

A third approach achieves some of the output benefits of tracking, with a lesser penalty in terms of land area, capital and operating cost. This involves tracking the sun in one dimension – in its daily journey across the sky – but not adjusting for the seasons. The angle of the axis is normally horizontal, though some, such as the solar park at Nellis Airforce Base, which have a 20° tilt, incline the axis towards the equator in a north-south orientation – effectively a hybrid between tracking and fixed tilt.

Single axis tracking systems are aligned along axes roughly North-South. Some use linkages between rows so that the same actuator can adjust the angle of several rows at once.

Power Conversion

Solar panels produce direct current (DC) electricity, so solar parks need conversion equipment to convert this to alternating current (AC), which is the form transmitted by the electricity grid. This conversion is done by inverters. To maximise their efficiency, solar power plants also incorporate maximum power point trackers, either within the inverters or as separate units. These devices keep each solar array string close to its peak power point.

There are two primary alternatives for configuring this conversion equipment; centralised and string inverters, although in some cases individual, or micro-inverters are used. Single inverters allows optimizing the output of each panel, and multiple inverters increases the reliability by limiting the loss of output when an inverter fails.

Centralised Inverters

These units have relatively high capacity, typically of the order of 1 MW, so they condition that the output of a substantial block of solar arrays, up to perhaps 2 hectares (4.9 acres) in area. Solar parks using centralised inverters are often configured in discrete rectangular blocks, with the related inverter in one corner, or the centre of the block.

Waldpolenz Solar Park is divided into blocks, each with a centralised inverter

String Inverters

String inverters are substantially lower in capacity, of the order of 10 kW, and condition the output of a single array string. This is normally a whole, or part of, a row of solar arrays within the overall plant. String inverters can enhance the efficiency of solar parks, where different parts of the array are experiencing different levels of insolation, for example where arranged at different orientations, or closely packed to minimise site area.

Transformers

The system inverters typically provide power output at voltages of the order of 480 V_{AC}. Electricity grids operate at much higher voltages of the order of tens or hundreds of thousands of volts, so transformers are incorporated to deliver the required output to the grid. Due to the long lead time, the Long Island Solar Farm chose to keep a spare transformer onsite, as transformer failure would have kept the solar farm offline for a long period. Transformers typically have a life of 25 to 75 years, and normally do not require replacement during the life of a photovoltaic power station.

System Performance

The performance of a solar park is a function of the climatic conditions, the equipment used and the system configuration. The primary energy input is the global light irradiance in the plane of the solar arrays, and this in turn is a combination of the direct and the diffuse radiation.

A key determinant of the output of the system is the conversion efficiency of the solar modules, which will depend in particular on the type of solar cell used.

There will be losses between the DC output of the solar modules and the AC power delivered to the grid, due to a wide range of factors such as light absorption losses, mismatch, cable voltage drop, conversion efficiencies, and other parasitic losses. A parameter called the 'performance ratio' has been developed to evaluate the total value of these losses. The performance ratio gives a measure of the output AC power delivered as a proportion of the total DC power which the solar modules should be able to deliver under the ambient climatic conditions. In modern solar parks the performance ratio should typically be in excess of 80%.

System Degradation

Early photovoltaic systems output decreased as much as 10%/year, but as of 2010 the median degradation rate was 0.5%/year, with modules made after 2000 having a significantly lower degradation rate, so that a system would lose only 12% of its output performance in 25 years. A system using modules which degrade 4%/year will lose 64% of its output during the same period. Many panel makers offer a performance guarantee, typically 90% in ten years and 80% over 25 years. The output of all panels is typically warranteed at plus or minus 3% during the first year of operation.

The Business of Developing Solar Parks

Solar power plants are developed to deliver merchant electricity into the grid as an alternative to other renewable, fossil or nuclear generating stations.

Westmill Solar Park is the world's largest community-owned solar power station

The plant owner is an electricity generator. Most solar power plants today are owned by independent power producers (IPP's), though some are held by investor- or community-owned utilities.

Some of these power producers develop their own portfolio of power plants, but most solar parks are initially designed and constructed by specialist project developers. Typically the developer will plan the project, obtain planning and connection consents, and arrange financing for the capital required. The actual construction work is normally contracted to one or more EPC (engineering, procurement and construction) contractors.

Major milestones in the development of a new photovoltaic power plant are planning consent, grid connection approval, financial close, construction, connection and commissioning. At each stage in the process, the developer will be able to update estimates of the anticipated performance and costs of the plant and the financial returns it should be able to deliver.

Planning Approval

Photovoltaic power stations occupy at least one hectare for each megawatt of rated output, so require a substantial land area; which is subject to planning approval. The chances of obtaining consent, and the related time, cost and conditions, varying from jurisdiction to jurisdiction and location to location. Many planning approvals will also apply conditions on the treatment of the site after the station has been decommissioned in the future. A professional health, safety and environment assessment is usually undertaken during the design of a PV power station in order to ensure the facility is designed and planned in accordance with all HSE regulations.

Grid Connection

The availability, locality and capacity of the connection to the grid is a major consideration in planning a new solar park, and can be a significant contributor to the cost.

Most stations are sited within a few kilometres of a suitable grid connection point. This network needs to be capable of absorbing the output of the solar park when operating at its maximum capacity. The project developer will normally have to absorb the cost of providing power lines to this point and making the connection; in addition often to any costs associated with upgrading the grid, so it can accommodate the output from the plant.

Operation and Maintenance

Once the solar park has been commissioned, the owner usually enters into a contract with a suitable counterparty to undertake operation and maintenance (O&M). In many cases this may be fulfilled by the original EPC contractor.

Solar plants' reliable solid-state systems require minimal maintenance, compared to rotating machinery for example. A major aspect of the O&M contract will be continuous monitoring of the performance of the plant and all of its primary subsystems, which is normally undertaken remotely. This enables performance to be compared with the anticipated output under the climatic conditions actually experienced. It also provides data to enable the scheduling of both rectification and preventive maintenance. A small number of large solar farms use a separate inverter or maximizer for each solar panel, which provide individual performance data that can be monitored. For other solar farms, thermal imaging is a tool that is used to identify non-performing panels for replacement.

Power Delivery

A solar park's income derives from the sales of electricity to the grid, and so its output is metered in real-time with readings of its energy output provided, typically on a half-hourly basis, for balancing and settlement within the electricity market.

Income is affected by the reliability of equipment within the plant and also by the availability of the grid network to which it is exporting. Some connection contracts allow the transmission system operator to constrain the output of a solar park, for example at times of low demand or high availability of other generators. Some countries make statutory provision for priority access to the grid for renewable generators, such as that under the European Renewable Energy Directive.

Economics and Finance

In recent years, PV technology has improved its electricity generating efficiency, reduced the installation cost per watt as well as its energy payback time (EPBT), and has reached grid parity in at least 19 different markets by 2014. Photovoltaics is increasingly becoming a viable source of mainstream power. However, prices for PV systems show strong regional variations, much more than solar cells and panels, which tend to be global commodities. In 2013, utility-scale system prices in highly penetrated markets such as China and Germany were significantly lower ($1.40/W) than in the United States ($3.30/W). The IEA explains these discrepancies due to differences in "soft costs", which include customer acquisition, permitting, inspection and interconnection, installation labor and financing costs.

Utility-scale PV system prices	
Country	**Cost ($/W)**
Australia	2.0
China	1.4
France	2.2
Germany	1.4

Italy	1.5
Japan	2.9
United Kingdom	1.9
United States	3.3
For utility-scale PV systems in 2013	

Grid Parity

Solar generating stations have become progressively cheaper in recent years, and this trend is expected to continue. Meanwhile, traditional electricity generation is becoming progressively more expensive. These trends are expected to lead to a crossover point when the levelised cost of energy from solar parks, historically more expensive, matches the cost of traditional electricity generation. This point is commonly referred to as grid parity.

For merchant solar power stations, where the electricity is being sold into the electricity transmission network, the levelised cost of solar energy will need to match the wholesale electricity price. This point is sometimes called 'wholesale grid parity' or 'busbar parity'.

Some photovoltaic systems, such as rooftop installations, can supply power directly to an electricity user. In these cases, the installation can be competitive when the output cost matches the price at which the user pays for his electricity consumption. This situation is sometimes called 'retail grid parity', 'socket parity' or 'dynamic grid parity'. Research carried out by UN-Energy in 2012 suggests areas of sunny countries with high electricity prices, such as Italy, Spain and Australia, and areas using diesel generators, have reached retail grid parity.

Incentive Mechanisms

Because the point of grid parity has not yet been reached in many parts of the world, solar generating stations need some form of financial incentive to compete for the supply of electricity. Many legislatures around the world have introduced such incentives to support the deployment of solar power stations.

Feed-in Tariffs

Feed in tariffs are designated prices which must be paid by utility companies for each kilowatt hour of renewable electricity produced by qualifying generators and fed into the grid. These tariffs normally represent a premium on wholesale electricity prices and offer a guaranteed revenue stream to help the power producer finance the project.

Renewable Portfolio Standards and Supplier Obligations

These standards are obligations on utility companies to source a proportion of their electricity from renewable generators. In most cases, they do not prescribe which technology should be used and the utility is free to select the most appropriate renewable sources.

There are some exceptions where solar technologies are allocated a proportion of the RPS in what is sometimes referred to as a 'solar set aside'.

Loan Guarantees and Other Capital Incentives

Some countries and states adopt less targeted financial incentives, available for a wide range of infrastructure investment, such as the US Department of Energy loan guarantee scheme, which stimulated a number of investments in the solar power plant in 2010 and 2011.

Tax Credits and Other Fiscal Incentives

Another form of indirect incentive which has been used to stimulate investment in solar power plant was tax credits available to investors. In some cases the credits were linked to the energy produced by the installations, such as the Production Tax Credits. In other cases the credits were related to the capital investment such as the Investment Tax Credits

International, National and Regional Programmes

In addition to free market commercial incentives, some countries and regions have specific programs to support the deployment of solar energy installations.

The European Union's Renewables Directive sets targets for increasing levels of deployment of renewable energy in all member states. Each has been required to develop a National Renewable Energy Action Plan showing how these targets would be met, and many of these have specific support measures for solar energy deployment. The directive also allows states to develop projects outside their national boundaries, and this may lead to bilateral programs such as the Helios project.

The Clean Development Mechanism of the UNFCCC is an international programme under which solar generating stations in certain qualifying countries can be supported.

Additionally many other countries have specific solar energy development programmes. Some examples are India's JNNSM, the Flagship Program in Australia, and similar projects in South Africa and Israel.

Financial Performance

The financial performance of the solar power plant is a function of its income and its costs.

The electrical output of a solar park will be related to the solar radiation, the capacity of the plant and its performance ratio. The income derived from this electrical output will come primarily from the sale of the electricity, and any incentive payments such as those under Feed-in Tariffs or other support mechanisms.

Electricity prices may vary at different times of day, giving a higher price at times of high demand. This may influence the design of the plant to increase its output at such times.

The dominant costs of solar power plants are the capital cost, and therefore any associated financing and depreciation. Though operating costs are typically relatively low, especially as no fuel is required, most operators will want to ensure that adequate operation and maintenance cover is available to maximise the availability of the plant and thereby optimise the income to cost ratio.

Geography

The first places to reach grid parity were those with high traditional electricity prices and high levels of solar radiation. Currently, more capacity is being installed in the rooftop than in the utility-scale segment. However, the worldwide distribution of solar parks is expected to change as different regions achieve grid parity. This transition also includes a shift from rooftop towards utility-scale plants, since the focus of new PV deployment has changed from Europe towards the *Sunbelt markets* where ground-mounted PV systems are favored.

Because of the economic background, large-scale systems are presently distributed where the support regimes have been the most consistent, or the most advantageous. Total capacity of worldwide PV plants above 4 MW_{AC} was assessed by Wiki-Solar as 36 GW in c. 2,300 installations at the end of 2014 and represents about 25 percent of total global PV capacity of 139 GW. The countries which had the most capacity, in descending order, were the United States, China, Germany, India, United Kingdom, Spain, Italy, Canada and South Africa. Activities in the key markets are reviewed individually below.

China

China was reported in early 2013 to have overtaken Germany as the nation with the most utility-scale solar capacity. Much of this has been supported by the Clean Development Mechanism. The distribution of power plants around the country is quite broad, with the highest concentration in the Gobi desert and connected to the Northwest China Power Grid.

Germany

The first multi-megawatt plant in Europe was the 4.2 MW community-owned project at Hemau, commissioned in 2003. But it was the revisions to the German feed-in tariffs in 2004, which gave the strongest impetus to the establishment of utility-scale solar power plants. The first to be completed under this programme was the Leipziger Land solar park developed by Geosol. Several dozen plants were built between 2004 and 2011, several of which were at the time the largest in the world. The EEG, the law which establishes Germany's feed-in tariffs, provides the legislative basis not just for the compensation levels, but other regulatory factors, such as priority access to the grid. The law was amended in 2010 to restrict the use of agricultural land, since which time most solar parks have been built on so-called 'development land', such as former military sites. Partly for this reason, the geographic distribution of photovoltaic power plants in Germany is biased towards the former Eastern Germany. As of February 2012, Germany had 1.1 million photovoltaic power plants (most are small kW roof mounted).

India

India has been rising up the leading nations for the installation of utility-scale solar capacity. The Charanka Solar Park in Gujarat was opened officially in April 2012 and was at the time the largest group of solar power plants in the world. Geographically the majority of the stations are located in Gujarat and Maharashtra. Rajasthan has successfully been attempting to attract solar development. It and Gujarat share the Thar Desert, along with Pakistan.

Italy

Italy has a very large number of photovoltaic power plants, the largest of which is the 84 MW Montalto di Castro project.

Spain

The majority of the deployment of solar power stations in Spain to date occurred during the boom market of 2007-8. The stations are well distributed around the country, with some concentration in Extremadura, Castile-La Mancha and Murcia.

United Kingdom

The introduction of Feed-in tariffs in the United Kingdom in 2010 stimulated the first wave of utility-scale projects, with c. 20 plants being completed before tariffs were reduced on 1 August 2011 following the 'Fast Track Review'. A second wave of installations was undertaken under the UK's Renewables Obligation, with the total number of plants connected by the end of March 2013 reaching 86. This is reported to have made the UK Europe's best market in the first quarter of 2013.

UK projects were originally concentrated in the South West, but have more recently spread across the South of England and into East Anglia and the Midlands. The first solar park in Wales came on stream in 2011 at Rhosygilwen, north Pembrokeshire. As of June 2014 there were 18 schemes generating more than 5 MW and 34 in planning or construction in Wales.

United States

The US deployment of photovoltaic power stations is largely concentrated in southwestern states. The Renewable Portfolio Standards in California and surrounding states provide a particular incentive. The volume of projects under construction in early 2013 has led to the forecast that the US will become the leading market.

Noteworthy Solar Parks

The following solar parks were, at the time they became operational, the largest in the world or their continent, or are notable for the reasons given:

Grid-connected Photovoltaic Power System

A grid-connected photovoltaic power system, or grid-connected PV power system is an electricity generating solar PV power system that is connected to the utility grid. A grid-connected PV system consists of solar panels, one or several inverters, a power conditioning unit and grid connection equipment. They range from small residential and commercial rooftop systems to large utility-scale solar power stations. Unlike stand-alone power systems, a grid-connected system rarely includes an integrated battery solution, as they are still very expensive. When conditions are right, the grid-connected PV system supplies the excess power, beyond consumption by the connected load, to the utility grid.

A grid-connected, residential solar rooftop system near Boston, USA

Operation

Residential, grid-connected rooftop systems which have a capacity more than 10 kilowatts can meet the load of most consumers. They can feed excess power to the grid where it is consumed by other users. The feedback is done through a meter to monitor power transferred. Photovoltaic wattage may be less than average consumption, in which case the consumer will continue to purchase grid energy, but a lesser amount than previously. If photovoltaic wattage substantially exceeds average consumption, the energy produced by the panels will be much in excess of the demand. In this case, the excess power can yield revenue by selling it to the grid. Depending on their agreement with their local grid energy company, the consumer only needs to pay the cost of electricity consumed less the value of electricity generated. This will be a negative number if more electricity is generated than consumed. Additionally, in some cases, cash incentives are paid from the grid operator to the consumer.

Photovoltaic power station at Nellis Air Force Base, United States

Connection of the photovoltaic power system can be done only through an interconnection agreement between the consumer and the utility company. The agreement details the various safety standards to be followed during the connection.

Features

Solar energy gathered by photovoltaic solar panels, intended for delivery to a power grid, must be conditioned, or processed for use, by a grid-connected inverter. Fundamentally, an inverter changes the DC input voltage from the PV to AC voltage for the grid. This inverter sits between the solar array and the grid, draws energy from each, and may be a large standalone unit or may be a collection of small inverters, each physically attached to individual solar panels. The inverter must monitor grid voltage, waveform, and frequency. One reason for monitoring is if the grid is dead or strays too far out of its nominal specifications, the inverter must not pass along any solar energy. An inverter connected to a malfunctioning power line will automatically disconnect in accordance with safety rules, for example UL1741, which vary by jurisdiction. Another reason for the inverter monitoring the grid is because for normal operation the inverter must synchronize with the grid waveform, and produce a voltage slightly higher than the grid itself, in order for energy to smoothly flow outward from the solar array.

Anti-islanding

Islanding is the condition in which a distributed generator continues to power a location even though power from the electric utility grid is no longer present. Islanding can be dangerous to utility workers, who may not realize that a circuit is still powered, even though there is no power from the electrical grid. For that reason, distributed generators must detect islanding and immediately stop producing power; this is referred to as anti-islanding.

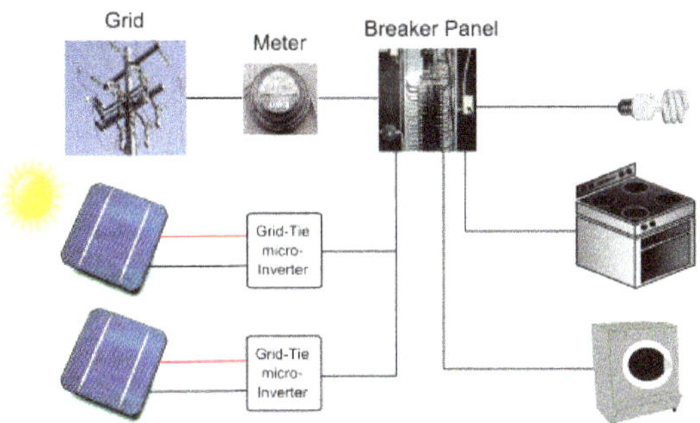

Diagram of a residential grid-connected PV system

In the case of a utility blackout in a grid-connected PV system, the solar panels will continue to deliver power as long as the sun is shining. In this case, the supply line becomes an "island" with power surrounded by a "sea" of unpowered lines. For this reason, solar inverters that are designed to supply power to the grid are generally required to have automatic anti-islanding circuitry in them. In intentional islanding, the generator disconnects from the grid, and forces the distributed generator to power the local circuit. This is often used as a power backup system for buildings that normally sell their power to the grid.

There are two types of anti-islanding control techniques:

- *Passive:* The voltage and/or the frequency change during the grid failure is measured and a positive feedback loop is employed to push the voltage and/or the frequency further away from its nominal value. Frequency or voltage may not change if the load matches very well with the inverter output or the load has a very high quality factor (reactive to real power ratio). So there exists some *Non Detection Zone* (NDZ).
- *Active:* This method employs injecting some error in frequency or voltage. When grid fails, the error accumulates and pushes the voltage and/or frequency beyond the acceptable range.

Advantages

- Systems such as Net Metering and Feed-in Tariff which are offered by some system operators, can offset a customers electricity usage costs. In some locations though, grid technologies cannot cope with distributed generation feeding into the grid, so the export of surplus electricity is not possible and that surplus is earthed.
- Grid-connected PV systems are comparatively easier to install as they do not require a battery system.

- Grid interconnection of photovoltaic (PV) power generation systems has the advantage of effective utilization of generated power because there are no storage losses involved.
- A photovoltaic power system is carbon negative over its lifespan, as any energy produced over and above that to build the panel initially offsets the need for burning fossil fuels. Even though the sun doesn't always shine, any installation gives a reasonably predictable average reduction in carbon consumption.

Disadvantages

- Grid-connected PV can cause issues with voltage regulation. The traditional grid operates under the assumption of one-way, or radial, flow. But electricity injected into the grid increases voltage, and can drive levels outside the acceptable bandwidth of ±5%.
- Grid-connected PV can compromise power quality. PV's intermittent nature means rapid changes in voltage. This not only wears out voltage regulators due to frequent adjusting, but also can result in voltage flicker.
- Connecting to the grid poses many protection-related challenges. In addition to islanding, as mentioned above, too high levels of grid-connected PV result in problems like relay desensitization, nuisance tripping, interference with automatic reclosers, and ferroresonance.

Rooftop Photovoltaic Power Station

A rooftop photovoltaic power station, or rooftop PV system, is a photovoltaic system that has its electricity-generating solar panels mounted on the rooftop of a residential or commercial building or structure. The various components of such a system include photovoltaic modules, mounting systems, cables, solar inverters and other electrical accessories.

Rooftop mounted systems are small compared to ground-mounted photovoltaic power stations with capacities in the megawatt range. Rooftop PV systems on residential buildings typically feature a capacity of about 5 to 20 kilowatts (kW), while those mounted on commercial buildings often reach 100 kilowatts or more.

Installation

The urban environment provides a large amount of empty rooftop spaces and can inherently avoid the potential land use and environmental concerns. Estimating rooftop solar insolation is a multi-faceted process, as insolation values in rooftops are impacted by the following:

Rooftop PV systems at Googleplex, California

- Time of the year
- Latitude
- Weather conditions
- Roof slope
- Roof aspect
- Shading from adjacent buildings and vegetation

There are various methods for calculating potential solar PV roof systems including the use of Lidar and orthophotos. Sophisticated models can even determine shading losses over large areas for PV deployment at the municipal level.

Feed-in Tariff Mechanism

In a grid connected rooftop photovoltaic power station, the generated electricity can be sold to the grid at a price higher than what the grid charges for the consumers. This arrangement provides payback for the investment of the installer. Many consumers from across the world are switching to this mechanism owing to the revenue yielded. The FIT as it is commonly known has led to an expansion in the solar PV industry worldwide. Thousands of jobs have been created through this form of subsidy. However it can produce a bubble effect which can burst when the FIT is removed. It has also increased the ability for localised production and embedded generation reducing transmission losses through power lines.

Hybrid Systems

A rooftop photovoltaic power station (either on-grid or off-grid) can be used in conjunction with other power sources like diesel generators, wind turbine etc. This system is capable of providing a continuous source of power.

Rooftop PV hybrid system.

Advantages

Installers have the right to feed solar electricity into the public grid and hence receive a reasonable premium tariff per generated kWh reflecting the benefits of solar electricity to compensate for the current extra costs of PV electricity.

Disadvantages

An electrical power system containing a 10% contribution from PV stations would require a 2.5% increase in load frequency control (LFC) capacity over a conventional system. The break-even cost for PV power generation is found to be relatively high for contribution levels of less than 10%. Higher proportions of PV power generation gives lower break-even costs, but economic and LFC considerations imposed an upper limit of about 10% on PV contributions to the overall power systems.

Technical Challenges

There are many technical challenges to integrating large amounts of rooftop PV systems to the power grid. For example:

- Reverse Power Flow

  The electric power grid was not designed for two way power flow at the distribution level. Distribution feeders are usually designed as a radial system for one way power flow transmitted over long distances from large centralized generators to customer loads at the end of the distribution feeder. Now with localized and distributed solar PV generation on rooftops, reverse flow causes power to flow to the substation and transformer, causing significant challenges. This has adverse effects on protection coordination and voltage regulators and protection coordination.

- Ramp rates

  Rapid fluctuations of generation from PV systems due to intermittent clouds cause undesirable levels of voltage variability in the distribution feeder. At high penetration of rooftop PV, this voltage variability reduces the stability of the grid due to transient imbalance in load and generation and causes voltage and frequency to exceed set limits. That is, the centralized generators cannot ramp fast enough to match the variability of the PV systems causing frequency mismatch on the whole system. This could lead to blackouts. This is an example of how a simple localized rooftop PV system can affect the whole power grid.

Cost

Residential PV system prices (2013)	
Country	**Cost ($/W)**
Australia	1.8
China	1.5
France	4.1
Germany	2.4
Italy	2.8
Japan	4.2
United Kingdom	2.8
United States	4.9
For residential PV systems in 2013	

Commercial PV system prices (2013)	
Country	**Cost ($/W)**
Australia	1.7
China	1.4
France	2.7
Germany	1.8
Italy	1.9
Japan	3.6
United Kingdom	2.4
United States	4.5
For commercial PV systems in 2013	

Future Prospects

The Jawaharlal Nehru National Solar Mission of the Indian government is planning to install utility scale grid-connected solar photovoltaic systems including rooftop photovoltaic systems with the combined capacity of up to 100 gigawatts by 2022.

Building-Integrated Photovoltaics

Building-integrated photovoltaics (BIPV) are photovoltaic materials that are used to replace conventional building materials in parts of the building envelope such as the roof, skylights, or facades. They are increasingly being incorporated into the construction of new buildings as a principal or ancillary source of electrical power, although existing buildings may be retrofitted with similar technology. The advantage of integrated photovoltaics over more common non-integrated systems is that the initial cost can be offset by reducing the amount spent on building materials and labor that would normally be used to construct the part of the building that the BIPV modules replace. These advantages make BIPV one of the fastest growing segments of the photovoltaic industry.

The CIS Tower in Manchester, England was clad in PV panels at a cost of £5.5 million. It started feeding electricity to the National Grid in November 2005.

The term building-applied photovoltaics (BAPV) is sometimes used to refer to photovoltaics that are a retrofit – integrated into the building after construction is complete. Most building-integrated installations are actually BAPV. Some manufacturers and builders differentiate new construction BIPV from BAPV.

History

PV applications for buildings began appearing in the 1970s. Aluminum-framed photovoltaic modules were connected to, or mounted on, buildings that were usually in remote areas without access to an electric power grid. In the 1980s photovoltaic module add-ons to roofs began being demonstrated. These PV systems were usually installed on utility-grid-connected buildings in areas with centralized power stations. In the 1990s

BIPV construction products specially designed to be integrated into a building envelope became commercially available. A 1998 doctoral thesis by Patrina Eiffert, entitled *An Economic Assessment of BIPV*, hypothesized that one day there would an economic value for trading Renewable Energy Credits (RECs). A 2011 economic assessment and brief overview of the history of BIPV by the U.S. National Renewable Energy Laboratory suggests that there may be significant technical challenges to overcome before the installed cost of BIPV is competitive with photovoltaic panels. However, there is a growing consensus that through their widespread commercialization, BIPV systems will become the backbone of the zero energy building (ZEB) European target for 2020. Despite technical promise, social barriers to widespread use have also been identified, such as the conservative culture of the building industry and integration with high-density urban design. These authors suggest enabling long-term use likely depends on effective public policy decisions as much as the technological development.

Photovoltaic wall near Barcelona, Spain

Projet BIPV, Gare de Perpignan, Southern France

PV Solar parking canopy, Autonomous University of Madrid, Spain

Forms

2009 Energy Project Award Winning 525 kilowatt BIPV CoolPly system manufactured by SolarFrameWorks, Co. on the Patriot Place Complex Adjacent to the Gillette Stadium in Foxborough, MA. System is installed on single-ply roofing membrane on a flat roof using no roof penetrations.

BAPV solar facade on a municipal building located in Madrid (Spain).

Building-Integrated Photovoltaic modules are available in several forms.

- Flat roofs
  - The most widely installed to date is an amorphous thin film solar cell integrated to a flexible polymer module which has been attached to the roofing membrane using an adhesive sheet between the solar module backsheet and the roofing membrane.
- Pitched roofs
  - Modules shaped like multiple roof tiles.
  - Solar shingles are modules designed to look and act like regular shingles, while incorporating a flexible thin film cell.
  - It extends normal roof life by protecting insulation and membranes from ultraviolet rays and water degradation. It does this by eliminating condensation because the dew point is kept above the roofing membrane.
- Facade
  - Facades can be installed on existing buildings, giving old buildings a whole new look. These modules are mounted on the facade of the build-

ing, over the existing structure, which can increase the appeal of the building and its resale value.

- Glazing
    - (Semi)transparent modules can be used to replace a number of architectural elements commonly made with glass or similar materials, such as windows and skylights.

What is BIPV?

- Three Main Types of BIPV Products
    - Crystalline silicon solar panels for ground-based and rooftop power plant
    - Amorphous crystalline silicon thin film solar pv modules which could be hollow, light, red blue yellow, as glass curtain wall and transparent skylight.
    - Double glasses solar panels with square cells inside.

Recently (in 2011-2016), Australian researchers have been working on developing the new approaches to implementing the unconventional BIPV systems and technologies, including solar photovoltaic windows of high visible transparency capable of providing significant energy savings due to superior thermal insulation properties and solar radiation control. Significant electric energy outputs were obtained from installation-ready framed PV window systems, as documented in

Transparent and Translucent Photovoltaics

Transparent solar panels use a tin oxide coating on the inner surface of the glass panes to conduct current out of the cell. The cell contains titanium oxide that is coated with a photoelectric dye.

Most conventional solar cells use visible and infrared light to generate electricity. In contrast, the innovative new solar cell also uses ultraviolet radiation. Used to replace conventional window glass, or placed over the glass, the installation surface area could be large, leading to potential uses that take advantage of the combined functions of power generation, lighting and temperature control.

Another name for transparent photovoltaics is "translucent photovoltaics" (they transmit half the light that falls on them). Similar to inorganic photovoltaics, organic photovoltaics are also capable of being translucent.

Government Subsidies

In some countries, additional incentives, or subsidies, are offered for building-integrated photovoltaics in addition to the existing feed-in tariffs for stand-alone solar systems. Since July 2006 France offered the highest incentive for BIPV, equal to an extra premium of EUR 0.25/kWh paid in addition to the 30 Euro cents for PV systems. These incentives are offered in the form of a rate paid for electricity fed to the grid.

European Union

- France €0.25/kWh
- Germany €0.05/kWh facade bonus expired in 2009
- Italy €0.04–€0.09/kWh
- United Kingdom 4.18 p/kWh
- Spain, compared with a non- building installation that receives €0.28/kWh (RD 1578/2008):
    - ≤20 kW: €0.34/kWh
    - >20 kW: €0.31/kWh

USA

- USA – Varies by state. Check Database of State Incentives for Renewables & Efficiency for more details.

China

Further to the announcement of a subsidy program for BIPV projects in March 2009 offering RMB20 per watt for BIPV systems and RMB15/watt for rooftop systems, the Chinese government recently unveiled a photovoltaic energy subsidy program "the Golden Sun Demonstration Project". The subsidy program aims at supporting the development of photovoltaic electricity generation ventures and the commercialization of PV technology. The Ministry of Finance, the Ministry of Science and Technology and the National Energy Bureau have jointly announced the details of the program in July 2009. Qualified on-grid photovoltaic electricity generation projects including rooftop, BIPV, and ground mounted systems are entitled to receive a subsidy equal to 50% of the total investment of each project, including associated transmission infrastructure. Qualified off-grid independent projects in remote areas will be eligible for subsidies of up to 70% of the total investment. In mid November, China's finance ministry has selected 294 projects totaling 642 megawatts that come to roughly RMB 20 billion ($3 billion) in costs for its subsidy plan to dramatically boost the country's solar energy production.

Other Integrated Photovoltaics

Vehicle-integrated photovoltaics (ViPV) are similar for vehicles. Solar cells could be embedded into panels exposed to sunlight such as the hood, roof and possibly the trunk depending on a car's design.

Concentrator Photovoltaics

Concentrator photovoltaics (CPV) is a photovoltaic technology that generates electricity from sunlight. Contrary to conventional photovoltaic systems, it uses lenses and curved mirrors to focus sunlight onto small, but highly efficient, multi-junction (MJ) solar cells. In addition, CPV systems often use solar trackers and sometimes a cooling system to further increase their efficiency. Ongoing research and development is rapidly improving their competitiveness in the utility-scale segment and in areas of high insolation. This sort of solar technology can be thus used in smaller areas.

This Amonix system consists of thousands of small lenses, each focusing sunlight to ~500X higher intensity onto a tiny, high-efficiency multi-junction solar cell. A Tesla Roadster is parked beneath for scale.

Concentrator photovoltaics (CPV) modules on dual axis solar trackers in Golmud, China

Systems using high concentrator photovoltaics (HCPV) especially have the potential to become competitive in the near future. They possess the highest efficiency of all existing PV technologies, and a smaller photovoltaic array also reduces the balance of system costs. Currently, CPV is not used in the PV rooftop segment and is far less common than conventional PV systems. For regions with a high annual direct normal irradiance of 2000 kilowatt-hour (kWh) per square meter or more, the levelized cost of electricity is in the range of \$0.08–\$0.15 per kWh and installation cost for a 10-megawatt CPV power plant was identified to lie between €1.40–€2.20 (~\$1.50-\$2.30) per watt-peak (W_p).

In 2013 CPV installations accounted for only 0.1%, or 50 megawatts (MW), of the annual global PV market of nearly 39,000 MW. However, by the end of 2014, cumulative installations already amounted to 330 MW. Commercial HCPV systems reached efficiencies of up to 42% with concentration levels above 400, and the International Energy Agency sees potential to increase the efficiency of this technology to 50% by the mid-2020s. As of December 2014, the best lab cell efficiency for concentrator MJ-cells reached 46% (four or more junctions). Most CPV installations are located in China, the United States, South Africa, Italy and Spain.

HCPV directly competes with concentrated solar power (CSP) as both technologies are suited best for areas with high direct normal irradiance, which are also known as the Sun Belt region in the United States and the Golden Banana in Southern Europe. CPV and CSP are often confused with one another, despite being intrinsically different technologies from the start: CPV uses the photovoltaic effect to directly generate electricity from sunlight, while CSP – often called *concentrated solar thermal* – uses the heat from the sun's radiation in order to make steam to drive a turbine, that then produces electricity using a generator. Currently, CSP is more common than CPV.

History

Research into concentrator photovoltaics has taken place since the 1970s. Sandia National Laboratories in Albuquerque, New Mexico was the site for most of the early work, with the first modern photovoltaic concentrating system produced there late in the decade. Their first system was a linear-trough concentrator system that used a point focus acrylic Fresnel lens focusing on water-cooled silicon cells and two axis tracking. Ramón Areces' system, also developed in the late 1970s, used hybrid silicone-glass Fresnel lenses, while cooling of silicon cells was achieved with a passive heat sink.

Challenges

CPV systems operate most efficiently in concentrated sunlight, as long as the solar cell is kept cool through use of heat sinks. Diffuse light, which occurs in cloudy and overcast conditions, cannot be concentrated. Filtered light, which occurs in hazy or polluted conditions, has spectral variations which can produce mismatches between the electrical currents generated within the series junctions of the spectrally-optimized MJ photovoltaic cells. To reach their maximum efficiency, CPV systems must be located in areas that receive plentiful direct, unfiltered sunlight.

The design of photovoltaic concentrators introduces a very specific optical design problem, with features that makes it different from any other optical design. It has to be efficient, suitable for mass production, capable of high concentration, insensitive to manufacturing and mounting inaccuracies, and capable of providing uniform illumination of the cell. All these reasons make nonimaging optics the most suitable for CPV.

CPV Strengths	CPV Weaknesses
High efficiencies for direct-normal irradiance	HCPV cannot utilize diffuse radiation. LCPV can only utilize a fraction of diffuse radiation
Low temperature coefficients	Tracking with sufficient accuracy and reliability is required
No cooling water required for passively cooled systems	May require frequent cleaning to mitigate soiling losses, depending on the site
Additional use of waste heat possible for systems with active cooling possible (e.g.large mirror systems)	Limited market – can only be used in regions with high DNI, cannot be easily installed on rooftops
Modular – kW to GW scale	Strong cost decrease of competing technologies for electricity production
Increased and stable energy production throughout the day due to tracking	Bankability and perception issues
Very low energy payback time	New generation technologies, without a history of production (thus increased risk)
Potential double use of land e.g. for agriculture, low environmental impact	Optical losses
High potential for cost reduction	Lack of technology standardization
Opportunities for local manufacturing	–
Smaller cell sizes could prevent large fluctuations in module price due to variations in semiconductor prices	–
Greater potential for efficiency increase in the future compared to single-junction flat plate systems could lead to greater improvements in land area use, BOS costs, and BOP costs	–
Source: Current Status of CPV report, January 2015. Table 2: Analysis of the strengths and weaknesses of CPV.	

Efficiency

All CPV systems have a concentrating optic and a solar cell. Except for very low concentrations, active solar tracking is also necessary. Low concentration systems often have a simple booster reflector, which can increase solar electric output by over 30% from that of non-concentrator PV systems. Experimental results from such LCPV systems in Canada resulted in energy gains over 40% for prismatic glass and 45% for traditional crystalline silicon PV modules.

Semiconductor properties allow solar cells to operate more efficiently in concentrated light, as long as the cell Junction temperature is kept cool by suitable heat sinks. Efficiency of multi-junction photovoltaic cells developed in research is upward of 44% today, with the potential to approach 50% in the coming years.

Also crucial to the efficiency (and cost) of a CPV system is the concentrating optic since it collects and concentrates sunlight onto the solar cell. For a given concentration, nonimaging optics combine the widest possible acceptance angles with high efficiency and, therefore, are the most appropriate for use in solar concentration. For very low concentrations, the wide acceptance angles of nonimaging optics avoid the need for active solar tracking. For medium and high concentrations, a wide acceptance angle can be seen as a measure of how tolerant the optic is to imperfections in the whole system. It is vital to start with a wide acceptance angle since it must be able to accommodate tracking errors, movements of the system due to wind, imperfectly manufactured optics, imperfectly assembled components, finite stiffness of the supporting structure or its deformation due to aging, among other factors. All of these reduce the initial acceptance angle and, after they are all factored in, the system must still be able to capture the finite angular aperture of sunlight.

Grid Parity

Grid parity refers to the cost of solar/wind watt-hours produced compared to watt-hours available from the electrical utility grid. Grid parity is achieved when renewable energy watt-hours are monetarily equal to watt-hours produced on the grid (from coal, hydro, etc.).

Compared to conventional flat panel solar cells, CPV might be advantageous because the solar collector is less expensive than an equivalent area of solar cells. However CPV hardware (solar collector and tracker) is nearing US$1 per watt, whereas silicon flat panels that are commonly sold are now below $1 per watt (not including any associated power systems or installation charges).

Types

CPV systems are categorized according to the amount of their solar concentration, measured in "suns" (the square of the magnification).

Low Concentration PV (LCPV)

An example of a Low Concentration PV Cell's surface, showing the glass lensing

Low concentration PV are systems with a solar concentration of 2–100 suns. For economic reasons, conventional or modified silicon solar cells are typically used, and, at these concentrations, the heat flux is low enough that the cells do not need to be actively cooled. There is now modeling and experimental evidence that standard solar modules do not need any modification, tracking or cooling if the concentration level is low and yet still have increased output of 35% or more. The laws of optics dictate that a solar collector with a low concentration ratio can have a high acceptance angle and thus in some instances does not require active solar tracking.

Medium Concentration PV

From concentrations of 100 to 300 suns, the CPV systems require two-axes solar tracking and cooling (whether passive or active), which makes them more complex.

A 10×10 mm HCPV solar cell

High Concentration Photovoltaics (HCPV)

High concentration photovoltaics (HCPV) systems employ concentrating optics consisting of dish reflectors or fresnel lenses that concentrate sunlight to intensities of 1,000 suns or more. The solar cells require high-capacity heat sinks to prevent thermal destruction and to manage temperature related electrical performance and life expectancy losses. To further exacerbate the concentrated cooling design, the heat sink must be passive, otherwise the power required for active cooling will reduce the overall conversion efficiency and economy. Multi-junction solar cells are currently favored over single junction cells, as they are more efficient and have a lower temperature coefficient (less loss in efficiency with an increase in temperature). The efficiency of both cell types rises with increased concentration; multi-junction efficiency rises faster. Multi-junction solar cells, originally designed for non-concentrating PV on space-based satellites, have been re-designed due to the high-current density encountered with CPV (typically 8 A/cm^2 at 500 suns). Though the cost of multi-junction solar cells is roughly 100 times that of conventional silicon cells of the same area, the small cell area employed makes the relative costs of cells in each system comparable and the system economics favor the multi-junction cells. Multi-junction cell efficiency has now reached 44% in production cells.

The 44% value given above is for a specific set of conditions known as "standard test conditions". These include a specific spectrum, an incident optical power of 850 W/m^2, and a cell temperature of 25 °C. In a concentrating system, the cell will typically operate under conditions of variable spectrum, lower optical power, and higher temperature. The optics needed to concentrate the light have limited efficiency themselves, in the range of 75–90%. Taking these factors into account, a solar module incorporating a 44% multi-junction cell might deliver a DC efficiency around 36%. Under similar conditions, a crystalline silicon module would deliver an efficiency of less than 18%.

When high concentration is needed (500–1000 times), as occurs in the case of high efficiency multi-junction solar cells, it is likely that it will be crucial for commercial success at the system level to achieve such concentration with a sufficient acceptance angle. This allows tolerance in mass production of all components, relaxes the module assembling and system installation, and decreasing the cost of structural elements. Since the main goal of CPV is to make solar energy inexpensive, there can be used only a few surfaces. Decreasing the number of elements and achieving high acceptance angle, can be relaxed optical and mechanical requirements, such as accuracy of the optical surfaces profiles, the module assembling, the installation, the supporting structure, etc. To this end, improvements in sunshape modelling at the system design stage may lead to higher system efficiencies.

Luminescent Solar Concentrators

A new emerging type of concentrators which are still at the research stage are Luminescent solar concentrators, they are composed of luminescent plates either totally impregnated by luminescent species or fluorescent thin films on transparent plates. They absorb solar light which is converted to fluorescence guided to plate edges where it emerges in a concentrated form. The concentration factor is directly proportional to the plate surface and inversely proportional to the plate edges. Such arrangement allows to use small amounts of solar cells as a result of concentration of fluorescent light. The fluorescent concentrator is able to concentrate both direct and diffuse light which is especially important on cloudy days. They also don't need expensive Solar trackers.

Installations

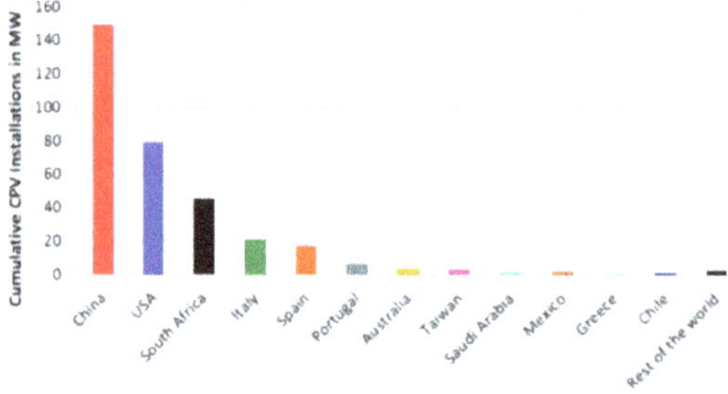

Cumulative CPV Installations in MW by country by November 2014

Concentrator photovoltaics technology has established its presence in the solar industry in the past few years. The first CPV power plant that exceeded 1 MW-level was commissioned in Spain in 2006. By the end of 2014, the fast-growing number CPV power plants around the world accounted for a total installed capacity of 330 MW.

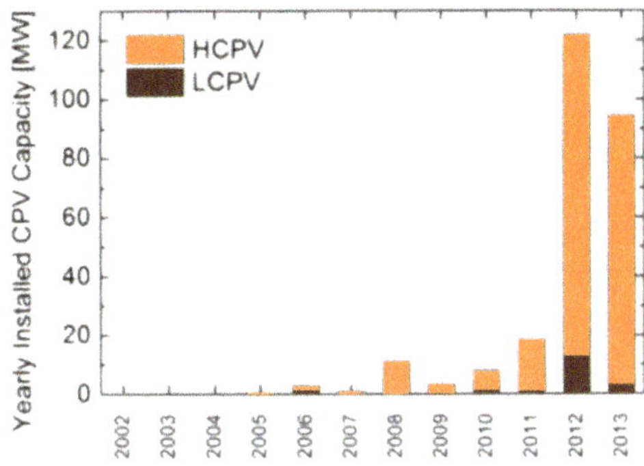

Yearly Installed CPV Capacity in MW from 2002 to 2013.

List of Large CPV Systems

The largest CPV power plant currently in operation is of 80 MW_p capacity located in Golmud, China, hosted by Suncore Photovoltaics.

Power station	Capacity (MW_p)	Location	Ref
Golmud 2	79.83	in Golmud/Qinghai province/China	
Golmud 1	57.96	in Golmud/Qinghai province/China	
Alamosa Solar Project	35.28	in Alamosa, Colorado/San Luis Valley/United States	
Source: The CPV Consortium			

Concentrated Photovoltaics and Thermal

Concentrator photovoltaics and thermal (CPVT), also sometimes called combined heat and power solar (CHAPS) or hybrid thermal CPV, is a cogeneration or micro cogeneration technology used in the field of concentrator photovoltaics that produces usable heat and electricity within the same system. CPVT at high concentrations of over 100 suns (HCPVT) utilizes similar components as HCPV, including dual-axis tracking and multi-junction photovoltaic cells. A fluid actively cools the integrated thermal–photovoltaic receiver, and simultaneously transports the collected heat.

Typically, one or more receivers and a heat exchanger operate within a closed thermal loop. To maintain efficient overall operation and avoid damage from thermal runaway, the demand for heat from the secondary side of the exchanger must be consistently high. Under such optimal operating conditions, collection efficiencies exceeding 70%

(up to ~35% electric, ~40% thermal for HCPVT) are anticipated. Net operating efficiencies may be substantially lower depending on how well a system is engineered to match the demands of the particular thermal application.

The maximum temperature of CPVT systems is typically too low alone to power a boiler for additional steam-based cogeneration of electricity. Such systems may be economical to power lower temperature applications having a constant high heat demand. The heat may be employed in district heating, water heating and air conditioning, desalination or process heat. For applications having lower or intermittent heat demand, a system may be augmented with a switchable heat dump to the external environment in order to maintain reliable electrical output and safeguard cell life, despite the resulting reduction in net operating efficiency.

HCPVT active cooling enables the use of much higher power thermal–photovoltaic receiver units, generating typically 1–100 kilowatts electric, as compared to HCPV systems that mostly rely upon passive cooling of single ~20W cells. Such high-power receivers utilize dense arrays of cells mounted on a high-efficiency heat sink. Minimizing the number of individual receiver units is a simplification that should ultimately yield improvement in the overall balance of system costs, manufacturability, maintainability/upgradeability, and reliability.

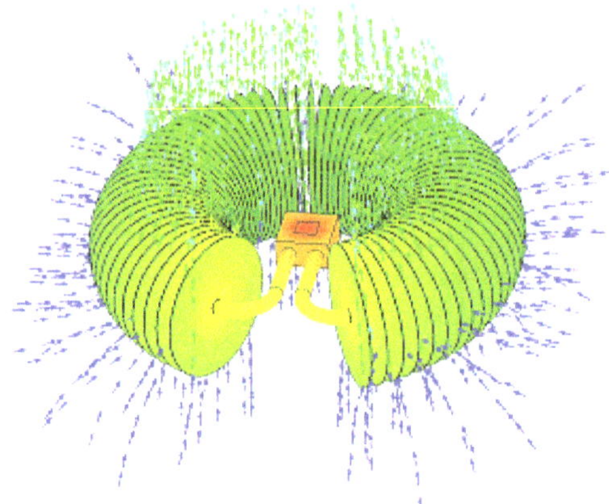

This 240 x 80 mm 1,000 suns CPV heat sink design thermal animation, was created using high resolution CFD analysis, and shows temperature contoured heat sink surface and flow trajectories as predicted.

Reliability Requirements

The maximum operating temperatures ($T_{max\ cell}$) of CPVT systems are limited to less than approximately 100–125 °C on account of the intrinsic reliability limitation of their multi-junction PV cells. This contrasts to CSP and other CHP systems which may be designed to function at temperatures in excess of several hundred degrees. More specifically, the multi-junction photovoltaic cells are fabricated from a layering of thin-film III-V semiconductor materials having intrinsic lifetimes during CPV operation that rapidly decrease with an Arrhenius-type temperature dependence. The system receiver must therefore provide for highly efficient and uniform cell cooling, where an ideal receiver would provide $T_{max\ coolant} \sim T_{max\ cell}$. In addition to material and design limitations in receiver heat-transfer performance, numerous extrinsic factors, such as the frequent system thermal cycling, further reduce the practical $T_{max\ coolant}$ compatible with long system life to below about 80 °C.

The higher capital costs, lesser standardization, and added engineering & operational complexities (in comparison to zero and low-concentration PV technologies) make demonstrations of system reliability and long-life performance critical challenges for the first generation of CPV and CPVT technologies. Performance certification testing standards (e.g. IEC 62108, UL 8703, IEC 62789, IEC 62670) include stress conditions that may be useful to uncover some predominantly infant and early life (<1–2 year) failure modes at the system, module, and sub-component levels. However, such standardized tests – as typically performed on only a small sampling of units – are generally incapable to evaluate comprehensive long-term (10 to 25 or more years) lifetimes for each unique CPVT system design and application under its broader range of actual operating conditions. Long-life performance of these complex systems is therefore assessed in the field, and is improved through aggressive product development cycles which are guided by the results of accelerated component/system aging, enhanced performance monitoring diagnostics, and failure analysis. Significant growth in the deployment of CPV and CPVT can be anticipated once the long-term performance and reliability concerns are better addressed to build confidence in system bankability.

Demonstration Projects

The economics of a mature CPVT industry is anticipated to be competitive, despite the large recent cost reductions and gradual efficiency improvements for conventional silicon PV (which can be installed alongside conventional CSP to provide for similar electrical+thermal generation capabilities). CPVT may currently be economical for niche markets having all of the following application characteristics:

- high solar direct normal incidence (DNI)
- tight space constraints for placement of a solar collector array
- high and constant demand for low-temperature (<80 °C) heat
- high cost of grid electricity
- access to backup sources of power or cost-efficient storage (electrical and thermal)

Utilization of a power purchase agreement (PPA), government assistance programs, and innovative financing schemes are also helping potential manufacturers and users to mitigate the risks of early CPVT technology adoption.

CPVT equipment offerings ranging from low (LCPVT) to high (HCPVT) concentration are now being deployed by several startup ventures. As such, longer-term viability of the technical and/or business approach being pursued by any individual system provider is typically speculative. Notably, the minimum viable products of startups can vary widely in their attention to reliability engineering. Nevertheless, the following incomplete compilation is offered to assist with the identification of some early industry trends.

LCPVT systems at ~14x concentration using reflective trough concentrators, and receiver pipes clad with silicon cells having dense interconnects, have been assembled by Cogenra with a claimed 75% efficiency (~15-20% electric, 60% thermal). Several such systems are in operation for more than 5 years as of 2015, and similar systems are being produced by Absolicon and Idhelio at 10x and 50x concentration, respectively.

HCPVT offerings at over 700x concentration have more recently emerged, and may be classified into three power tiers. Third tier systems are distributed generators consisting of large arrays of ~20W single-cell receiver/collector units, similar to those previously pioneered by Amonix and SolFocus for HCPV. Second tier systems utilize localized dense-arrays of cells that produce 1-100 kW of electrical power output per receiver/generator unit. First tier systems exceed 100 kW of electrical output and are most aggressive in targeting the utility market.

Several HCPVT system providers are listed in the following table. Nearly all are early demonstration systems which have been in service for under 5 years as of 2015. Collected thermal power is typically 1.5x-2x the rated electrical power.

Provider	Country	Concentrator Type	Unit Size in kW_e		Ref
			Generator	Receiver	
		- Tier 1 -			
Raygen	Australia	Large Heliostat Array	200	200	
Sunfish	United Kingdom	Large Heliostat Array	na	na	
		- Tier 2 -			
Renovalia	Spain	Large Dish	3	3	
Zenith Solar/Suncore	Israel/China	Large Dish	4.5	2.25	
Sun Oyster	Germany	Large Trough + Lens	4.7	2.35	
Forbes Solar	India/Germany	Large Dish	7.5	3.75	
Airlight Energy/dsolar	Switzerland	Large Dish	12	12	
Southwest Solar	United States	Large Dish	20	20	
Silex Solar Systems	Australia	Large Dish	35	35	

		- *Tier 3* -			
Brightleaf	United States	Small Dish Array	4	0.02	
Silex Power	Malta	Small Dish Array	16	0.04	
Solergy	Italy/USA	Small Lens Array	20	0.02	

Growth of Photovoltaics

Worldwide growth of photovoltaics has been fitting an exponential curve for more than two decades. During this period of time, photovoltaics (PV), also known as solar PV, has evolved from a pure niche market of small scale applications towards becoming a mainstream electricity source. When solar PV systems were first recognized as a promising renewable energy technology, programs, such as feed-in tariffs, were implemented by a number of governments in order to provide economic incentives for investments. For several years, growth was mainly driven by Japan and pioneering European countries. As a consequence, cost of solar declined significantly due to improvements in technology and economies of scale, even more so when production of solar cells and modules started to ramp up in China. Since then, deployment of photovoltaics is gaining momentum on a worldwide scale, particularly in Asia but also in North America and other regions, where solar PV is now increasingly competing with conventional energy sources as grid parity has already been reached in about 30 countries.

Projections for photovoltaic growth are difficult and burdened with many uncertainties. Official agencies, such as the International Energy Agency consistently increased their estimates over the years, but still fell short of actual deployment.

Historically, the United States had been the leader of installed photovoltaics for many years, and its total capacity amounted to 77 megawatts in 1996—more than any other country in the world at the time. Then, Japan stayed ahead as the world's leader of produced solar electricity until 2005, when Germany took the lead. The country is currently approaching the 40,000 megawatt mark. China is expected to continue its rapid growth and to triple its PV capacity to 70,000 megawatts by 2017. In 2015, China became world's largest producer of photovoltaic power. By the end of 2014, cumulative photovoltaic capacity reached at least 178 gigawatts (GW), sufficient to supply 1 percent of global electricity demands. Solar now contributes 7.9 percent and 7.0 percent to the respective annual domestic consumption in Italy and Germany. For 2015, worldwide deployment of about 55 GW is being forecasted, and installed capacity is projected to more than double or even triple beyond 500 GW between now and 2020. By 2050, solar power is anticipated to become the world's largest source of electricity, with solar photovoltaics and concentrated solar power contributing 16 and 11 percent, respectively. This will require PV capacity to grow to 4,600 GW, of which more than half is forecasted to be deployed in China and India.

Current Status

Current status describes worldwide, regional and domestic solar PV deployment as of the end of 2014. The unit of power, watt, is frequently used as multiples, such as kilowatt (kW), megawatt (MW), gigawatt (GW) and terawatt (TW). Nameplate capacity in the article is displayed as MW and has to be understood as direct current megawatt-peak (MW_p), if not otherwise explicitly denoted as, for example, MW_{AC}

Worldwide

In 2014, cumulative photovoltaic capacity increased by 40.1 GW or 28% and reached at least 178 GW by the end of the year, sufficient to supply 1 percent of the world's total electricity consumption of currently 18,400 TWh. Although this represents a new all-time record in the history of global PV deployment, overall expectations had been higher as module shipments amounted to 44–46 GW and suggested higher overall installations. Annual installation for 2014 expanded slightly by 5% when compared to worldwide installation of 38.3 GW in 2013.

Global installed PV capacity						
Report	**Cumulative (MW_p)**	**Installed (MW_p)**	**Year-End-Period**	**Release date**	**Type**	**Ref**
IEA-PVPS snap-shot	>177,000	>38,700	2014	March 2015	pre-liminary	
SPE outlook[(a)]	178,391	40,134	2014	June 2015	de-tailed	
IEA-PVPS trends	>177,000	~40,000	2014	October 2015	final	

Overview of reported annual installations and global cumulative figures in chronological order [(a)] SPE also noted that global PV-shipment was much higher, about 44–46 GW in 2014.

Regions

In 2014, Asia was the fastest growing region, with more than 60% of global installations. China and Japan alone accounted for 20 GW or half of worldwide deployment. Europe continued to decline and installed 7 GW or 18% of the global PV market, three times less than in the record-year of 2011, when 22 GW had been installed. For the first time, North and South America combined accounted for at least as much as Europe, about 7.1 GW or about 18% of global total. This is due to the strong growth in the United States, supported by Canada, Chile and Mexico.

In terms of cumulative capacity, Europe is still the most developed region with 88 GW or half of the global total of 178 GW. Solar PV now covers 3.5% and 7% of European electricity demand and peak electricity demand, respectively. The Asia-Pacific region (APAC) which includes countries such as Japan, India and Australia, follows second and accounts for about 20% percent of worldwide capacity. In third position ranks China with 16%, followed by the Americas with about 12%. Cumulative capacity in the MEA (Middle East and Africa) region and ROW (rest of the world) accounted for only about 3.3% of the global total. A great untapped potential remains for many of these countries, especially in the Sunbelt.

Countries

Chile (+0.4 GW) and South Africa (+0.8 GW) were the newcomers of 2014. South Africa entered the top 10 in added capacity rankings for the first time. There are now twenty countries around the world with a cumulative PV capacity of more than one gigawatt. Thailand (1,299 MW), The Netherlands (1,123 MW), and Switzerland (1,076 MW), all crossed the gigawatt threshold in 2014. Based on IEA's data, the available solar PV capacity in Italy, Germany and Greece is now sufficient to supply between 7% and 8% of their respective domestic electricity consumption.

Other mentionable PV deployments above the 100-megawatt mark included Canada (500 MW), Thailand (475 MW), The Netherlands (400 MW), Taiwan (400 MW), Italy (385 MW), Chile (1,113 MW), Switzerland (320 MW), Israel (250 MW), Austria (140 MW) and Portugal (110 MW). Underperforming countries were Belgium (65 MW), Bulgaria (2 MW), the Czech Republic (2 MW), Greece (16 MW), Romania (69 MW), Slovakia (0.4 MW) and Spain (22 MW), with very low to almost non-existent additions compared to previous years.

Top PV countries in 2015 (MW)				
Rank	Country	Total capacity	Date	Reference
1.	China	43,060	2015	
2.	Germany	39,640	2015	
3.	Japan	33,300	2015	
4.	United States	27,320	2015	
5.	Italy	18,920	2015	
6.	UK	9,080	2015	
7.	France	6,550	2015	
8.	India	5,170	2015	
9.	Spain	4,832	2015	

10.	Australia	4,100	2015	
11.	Belgium	3,200	2015	
12.	South Korea	3,200	2015	
13.	Greece	2,600	2015	
14.	Canada	2,240	2015	
15.	Czech Republic	2,070	2015	
16.	Thailand	1,600	2015	
17.	Switzerland	1,376	2015	
18.	South Africa	1,361	2015	
19.	Romania	1,301	2015	
20.	Netherlands	1,288	2015	
21.	Bulgaria	1,040	2015	
22.	Austria	900	2015	
23.	Chile	848	2015	
24.	Taiwan	800	2015	
25.	Denmark	791	2015	
26.	Israel	766	2015	
27.	Slovakia	533	2015	
28.	Honduras	455	2015	
29.	Portugal	454	2015	
30.	Ukraine	432	2015	
31.	Russia	407	2015	
32.	Algeria	274	2015	
33.	Turkey	249	2015	
34.	Slovenia	240	2015	
35.	Mexico	234	2015	
36.	Pakistan	210	2015	
37.	Malaysia	184	2015	
38.	Bangladesh	167	2015	
39.	Philippines	132	2015	
40.	Luxembourg	120	2015	

Forecast

Forecast for 2016

In April 2016, Mercom Capital Group, forecasted global solar installations to reach 66.7 GW with China, the United States, Japan and India to make up the top four solar markets in 2016.

Market research firm IHS forecast in February global installations to reach 69 GW in 2016. Meanwhile, Greentech Media has a 2016 forecast of 64 GW.

Forecast for 2015

In June 2015, SolarPower Europe (SPE) – the former European Photovoltaic Industry Association (EPIA) – released its new report, *Global Market Outlook for Solar Power 2015–2019*. The European PV organization expects global installations to grow between 41 GW and 60 GW, marked by their low and high scenario, respectively. A year before, the European lobby association estimated 2015 to grow by 35–53 GW.

The International Energy Agency (IEA) will still have to update its forecast in the course of 2015. This is expected to happen on 1 October 2015, when the *Medium-Term Renewable Energy Market Report 2015* will be launched on the sidelines of the G20 summit in Istanbul, Turkey. In August 2014, IEA forecasted 38 GW in its baseline scenario for 2015. The IEA has been criticized for systematically underestimating the growth of photovoltaics in the past.

IHS Technology forecasts global solar PV installations to grow by 59 GW or 33% in 2015. The company also predicts an accelerated growth for concentrator photovoltaics, an increase in market-share of monocrystalline silicon technology over polycrystalline silicon, currently the leading semiconductor material used for solar cells, and that solar power in California will provide more than 10 percent of the state's annual power generation, higher than in Italy and Germany.

Summary forecasts	
Projections for 2015	
IEA[(a)]	38 GW
SPE	51 GW
DB	54 GW
GTM	55 GW
BNEF	55 GW

MC	58 GW
IHS	59 GW
Ø for 2015	**55 GW**
(a) excluded from average	

Deutsche Bank (DB) anticipates deployment to reach about 54 GW in 2015. An increase in investments and improvement of cost competitiveness is expected, while weaker oil prices are not seen to play a significant role for the solar sector. They find that grid parity has arrived in 30 countries around the world (compared to 19 markets the year before), as unsubsidized rooftop solar costs $0.08–$0.13 per kilowatt-hour, and is now below the retail prices of electricity in these markets. DB also estimates current installation cost to range from $1.00/W for utility-scale systems in China to $2.90/W for U.S. residential rooftop systems.

For 2015, Mercom Capital (MC) predicts global installation to amount to 57.8 GW (up from 54.5 GW predicted a few months earlier), while Bloomberg New Energy Finance (BNEF) foresees solar PV to add more than 55 GW. The *10 Predictions For Clean Energy In 2015* by Michael Liebreich mentions the spread of PV to more and more localities in Africa, the trend of imposing taxes on rooftop systems, and the growing confidence among investors that solar is indeed a cheap source of power. In June 2015, Greentech Media (GTM) Research estimated global PV installations at 55 GW for 2015 and notes that this corresponds to about half of the world's newly installed electricity generating capacity.

About 40 countries are expected to install more than 100 megawatts in 2015 (compared to 25 countries in 2014). In the United States, installations are predicted to grow by 7.9 GW (SEIA) to 9.4 GW, up by about 30–45% over the record-year of 2014. Both, the United Kingdom (2.9–3.5 GW) and Japan (9–10.4 GW) are being forecasted to set new records in 2015. After three years of decline, installations in Europe are expected to grow again to 9.4 GW, up 19% over 2014. The Chinese government set its own 2015 solar target to 17.8 GW, much higher than its original 2014 target it ultimately missed to achieve. India is expected to install more than 2 GW, a tripling over the previous year. A return of deployment in the gigawatt-scale is predicted for France, and record installations of 1.1 GW are expected for Thailand, while deployment in Australia and Germany would remain unchanged. Latin America is forecasted to install 2.2 GW in 2015, with a significant contribution from the Central American region for the first time, while Chile and Mexico are expected to double and triple their installations, respectively. The projected top five Latin American installers of 2015 are Chile (1 GW), Honduras (460 MW), Mexico (195 MW), Guatemala (98 MW) and Panama (62 MW). Rapid growth of solar PV is also expected to occur in Jordan, Pakistan and the Philippines.

Global Short-term Forecast (2020)

Summary of forecasts			
Forecasting company or organization	**Cumulative by 2020**	**To be added 2015–2020[7]**	**Ø Annualinstallation**
IEA (baseline, 2014)	403 GW	225 GW	38 GW
GlobalData (2014)	414 GW	236 GW	39 GW
SPE/EPIA (low scenario, 2015)[1]	444 GW	266 GW	44 GW
Frost & Sullivan (2015)	446 GW	268 GW	45 GW
IEA (enhanced case, 2014)[2]	490 GW	312 GW	52 GW
Grand View Research (2015)	490 GW	312 GW	52 GW
Citigroup (CitiResearch, 2013)	500 GW	322 GW	54 GW
PVMA (medium scenario, 2015)[3]	536 GW	358 GW	60 GW
IHS (10.5% CAGR, 2015)[4]	566 GW	388 GW	65 GW
BNEF (New Energy Outlook 2015)[5]	589 GW	411 GW	69 GW
SPE/EPIA (high scenario, 2015)[1]	630 GW	452 GW	75 GW
Fraunhofer (17% CAGR, 2015)[6]	668 GW	490 GW	82 GW
GTM Research (June, 2015)	696 GW	518 GW	86 GW

List ordered by ascending estimated capacities and publication date [1] SPE – extrapolated 2019-projection (396 GW and 540 GW, resp.)

[2] IEA – arithmetic mean of 465–515 GW

[3] PVMA – average of scenarios (444–630 GW), read from diagram

[4] IHS – extrapolated 2019-estimate, based on a CAGR of 10.5%

[5] BNEF – figures may include contribution from CSP

[6] FSH – external expert scenario based on a CAGR of 17%

[7] Difference to global cumulative as of the end of 2014 (178 GW)

There are a number of short and medium term forecasts published by several organizations and market research companies. In addition, the International Energy Agency (IEA) and Solar Power Europe (SPE, the former EPIA) produce more than one scenario each. The summary table shows the different forecasts for global PV capacity by 2020. Projections are listed by ascending cumulative capacity. The table also shows the capacity that has to be installed from 2015–2020 and the average annual installation required to meet the projection. Conservative scenarios forecast capacity to reach 400 or more gigawatts, assuming declining annual installations from current levels, while more optimistic scenarios project cumulative capacity to grow beyond 500 GW. Only the most optimistic projections around 600 GW foresee annual installations to grow above 10 percent (p.a) in the near future.

The European Photovoltaic Industry Association expects the fastest PV growth to continue in China, South-East Asia, Latin America, the Middle-East, North Africa, and India. By 2019, worldwide capacity is projected to reach between 396 GW (low scenario) and 540 GW (high scenario). This corresponds to a more than doubling and tripling of installed capacity within five years, respectively.

Consulting firm Frost & Sullivan projects global PV capacity to increase to 446 GW by 2020, with China, India and North America being the fastest growing regions, while Europe is expected to double its solar capacity from current levels. Grand View Research, a market research and consulting firm, headquartered in San Francisco, published its solar PV forecast report in March 2015. The large PV potential in countries such as Brazil, Chile and Saudi Arabia has not expanded as expected and is supposed to be explored over the next six years. In addition, China's increase of manufacturing capacity is expected to further lowering global market prices. The consulting firm projects worldwide cumulative deployment to reach about 490 GW by 2020.

The *PV Market Alliance* (PVMA), a recently founded consortium of several research bodies, sees global PV capacity to reach 444–630 GW by 2020. In its low scenario, annual installations are projected to grow from 40 to 50 gigawatts by the end of the decade, while its high scenario forecasts deployment to increase from 60 to 90 GW during the next five years. The medium scenario therefore expects annual PV installations to grow from 50 GW to 70 GW and to reach 536 GW by 2020. PVMA's figures are in line with those published earlier by Solar Power Europe. In June 2015, Greentech Media (GTM) Research released its *Global PV Demand Outlook* for 2020. The company projects annual installations to increase from 40 GW to 135 GW and global cumulative capacity to reach almost 700 GW by 2020. GTM's outlook is the most aggressive of all forecasts to date, with projected deployment of 518 GW between 2015 and 2020, or more than twice as much as IEA's 225 GW baseline case scenario, published ten months earlier.

IEA – projected annual PV installations				
Year	**2013-edition**	**Diff**	**2014-edition**	**2015-edition**
2013	30 GW	**+9**	39 GW	-
2014	30 GW	**+9**	39 GW	39 GW
2015	33 GW	**+5**	38 GW	42 GW
2016	36 GW	**+3**	39 GW	42 GW
2017	38 GW	**-2**	36 GW	39 GW
2018	40 GW	**-3**	37 GW	41 GW
2019	n.a.	n.a.	38 GW	43 GW
2020	n.a.	n.a.	39 GW	45 GW
Source: IEA *Medium Term Renewable Energy Market* data from 2013-edition, 2014-edition and 2015-edition				

The International Energy Agency (IEA) sees overall stagnating annual installations in the range of 36–39 GW until 2020, when global capacity will reach 403 GW, according to the highlighted baseline case scenario of the *Medium Term Renewable Energy Market 2014* report. Paradoxically, since the report's 2013-edition, projected cumulative for 2018 has increased by 6% from 308 GW to 326 GW, while the corresponding annual deployment decreased. This is due to the fact that the International Energy Agency adjusted annual installations upward on the near-end – in order to meet actual deployment, while reducing estimates on the far-end. The result is a flat curve that stays below 40 GW until 2020 *(see table)*. For 2017, the projected low of 36 GW concurs with the scheduled expiration of the solar investment tax credit (ITC) in the U.S. and the expected end of the solar boom in Japan. IEA's projected annual installation of less than 40 GW also leads to a negative growth rate, since expectations for 2015 are much higher. Such a decline, however, is unprecedented and has never been observed in the recorded history of solar PV deployment. This scenario makes IEA's baseline case the most conservative of all projections. In the less featured *enhanced high case scenario*, IEA estimates that "solar PV could reach a cumulative 465 GW to 515 GW in 2020" and that "solar PV capacity could top 500 GW globally in 2020".

By 2020, IEA's *Technology Roadmap: Solar Photovoltaic Energy report* expects China to account for over 110 GW of solar PV, while Japan and Germany would each reach around 50 GW. The United States would rank fourth at over 40 GW, followed by Italy and India with 25 GW and 15 GW. The United Kingdom, France and Australia, would have installed capacities of close to 10 GW each. IEA released this outlook in September 2014. Two months later, however, India announced its intention to install 100 GW of solar PV by 2022, and another six months later, SEIA forecasted that the United States would reach 40 GW of cumulative PV capacity already by the end of 2016. Furthermore, in July 2015, the UK was forecast to reach 10 GW by early 2016. IEA will release its next roadmap report on solar PV in 2018.

Global Long-term Forecast (2050)

In 2014, the International Energy Agency (IEA) released its latest edition of the *Technology Roadmap: Solar Photovoltaic Energy* report, calling for clear, credible and consistent signals from policy makers. The IEA also acknowledged to have previously underestimated PV deployment and reassessed its short-term and long-term goals.

IEA report Technology Roadmap: Solar Photovoltaic Energy (September 2014) –

> Much has happened since our 2010 IEA technology roadmap for PV energy. PV has been deployed faster than anticipated and by 2020 will probably reach twice the level previously expected. Rapid deployment and falling costs have each been driving the other. This progress, together with other important changes in the energy landscape, notably concerning the status and progress of nuclear power and CCS, have led the IEA to reassess the role of solar PV in mitigating climate change. This updated roadmap envisions PV's share of global electricity rising up to 16% by 2050, compared with 11% in the 2010 roadmap.

IEA's long-term scenario for 2050 describes worldwide solar photovoltaics (PV) and concentrated solar thermal (CSP) capacity to reach 4,600 GW and 1,000 GW, respectively. In order to achieve IEA's projection, PV deployment of 124 GW and investments of $225 billion are required annually. This is about three and two times of current levels, respectively. By 2050, levelized cost of electricity (LCOE) generated by solar PV would cost between US 4¢ and 16¢ per kilowatt-hour (kWh), or by segment and on average, 5.6¢ per kWh for utility-scale power plants (range of 4¢ to 9.7¢), and 7.8¢ per kWh for solar rooftop systems (range of 4.9¢ to 15.9¢) These estimates are based on a weighted average cost of capital (WACC) of 8%. The report notes that when the WACC exceeds 9%, over half the LCOE of PV is made of financial expenditures, and that more optimistic assumptions of a lower WACC would therefore significantly reduce the LCOE of solar PV in the future. The IEA also emphasizes that these new figures are not projections but rather scenarios they believe would occur if underlying economic, regulatory and political conditions played out.

In 2015, Fraunhofer ISE did a study commissioned by German renewable think tank *Agora Energiewende* and concluded that most scenarios fundamentally underestimate the role of solar power in future energy systems. Fraunhofer's study *(see summary of its conclusions below)* differs significantly form IEA's roadmap report on solar PV technology despite being published only a few months apart. The report foresees worldwide installed PV capacity to reach as much as 30,700 GW by 2050. By then, Fraunhofer expects LCOE for utility-scale solar farms to reach €0.02 to €0.04 per kilowatt-hour, or about half of what the International Energy Agency has been projecting (4¢ to 9.7¢). Turnkey system costs would decrease by more than 50% to €436/kW_p from currently €995/kW_p. This is also noteworthy, as IEA's roadmap published significantly higher estimates of $1,400 to $3,300 per kW_p for eight major markets around the world *(see table Typical PV system prices in 2013 below)*. However, the study agrees with IEA's roadmap report by emphasizing the importance of the cost of capital (WACC), which strongly depends on regulatory regimes and may even outweigh local advantages of higher solar insolation. In the study, a WACC of 5%, 7.5% and 10% is used to calculate the projected levelized cost of electricity for utility-scale solar PV in 18 different markets worldwide.

Fraunhofer ISE: Current and Future Cost of Photovoltaics. Long-term Scenarios for Market Development, System Prices and LCOE of Utility-Scale PV Systems. Study on behalf of Agora Energiewende (February 2015) –

1. Solar photovoltaics is already today a low-cost renewable energy technology. Cost of power from large scale photovoltaic installations in Germany fell from over 40 ct/kWh in 2005 to 9 cts/kWh in 2014. Even lower prices have been reported in sunnier regions of the world, since a major share of cost components is traded on global markets.

2. Solar power will soon be the cheapest form of electricity in many regions of the world. Even in conservative scenarios and assuming no major technological

breakthroughs, an end to cost reduction is not in sight. Depending on annual sunshine, power cost of 4–6 cts/kWh are expected by 2025, reaching 2–4 ct/kWh by 2050 (conservative estimate).

3. Financial and regulatory environments will be key to reducing cost in the future. Cost of hardware sourced from global markets will decrease irrespective of local conditions. However, inadequate regulatory regimes may increase cost of power by up to 50 percent through higher cost of finance. This may even overcompensate the effect of better local solar resources.

4. Most scenarios fundamentally underestimate the role of solar power in future energy systems. Based on outdated cost estimates, most scenarios modeling future domestic, regional or global power systems foresee only a small contribution of solar power. The results of our analysis indicate that a fundamental review of cost-optimal power system pathways is necessary.

Regional Forecasts

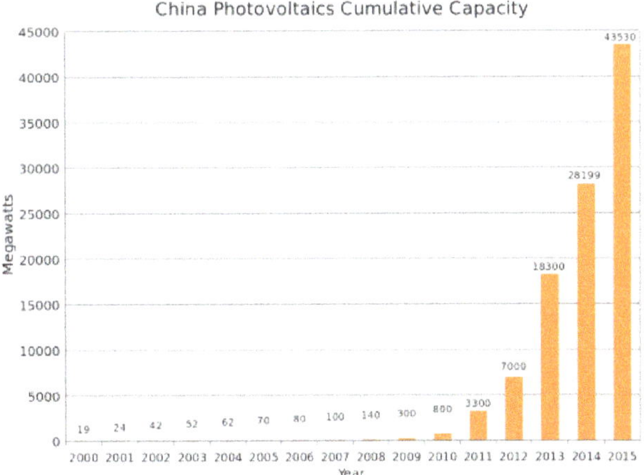

PV capacity growth in China

- China

In October 2015, China's National Energy Administration (NDRC) set an ambitious 23.1 GW target for 2015, upgrading its previous target of 17.8 GW from March 2015, which was already more than the entire global PV capacity installed in 2010. With this revised target, China will have surpassed Germany's total capacity of 40 GW by the end of the year and become the world's largest overall producer of photovoltaic power.

As of October 2015, China plans to install 150 GW of solar power by 2020, an increase of 50 GW compared to the 2020-target announced in October 2014, when China planned to install 100 GW of solar power—along with 200 GW of wind, 350 GW of hydro and 58 GW of nuclear power.

Overall, China has consistently increased its annual and short term targets. However estimates, targets and actual deployment have differed substantially in the past: in 2013 and 2014, China was expected to continue to install 10 GW per year. In February 2014, China's NDRC upgraded its 2014 target from 10 GW to 14 GW (later adjusted to 13 GW) and ended up installing an estimated 10.6 GW due to shortcomings in the distributed PV sector.

- India

By 2016, India plans to have constructed the world's largest solar farm with a capacity of 750 MW. The country plans to install 100 GW capacity of solar power by 2022, a five-time increase from a previous target. However, India's solar PV deployment was below expectation and actually declined from 1,115 MW in 2013 to 616 MW in 2014, which contrasts with the country's positive policy tone towards solar PV. In 2015, record installations are 3018 MW against expectation of 2 GW. As of July 13, 2015, total commissioned capacity in India surpassed 4 GW and stands at 4,096 MW, with the state Rajasthan and Gujarat taking the lead with 1,164 MW and 1,000 MW, respectively.

- Japan

For 2014, installations in Japan reached an all time record level of 9.7 GW, compared to 6.9 GW in 2013. By the end of 2014, Japan's installed PV capacity of 23.3 GW can now contribute about 2.5% to the overall domestic electricity demand. In 2014, Japan also overtook Italy (18.5 GW) as the world's third largest PV nation in terms of cumulative capacity. IHS forecasts that Japan will retain its position as the world's second largest solar market for new installations and grow by 4%, adding another 10.4 GW in 2015.

- United States

In March 2015, SEIA, the Solar Energy Industries Association and GTM Research, released their U.S. estimate for 2014. In the United States, a total of 6.2 gigawatts had been installed, up 30 percent over 2013 (vs. previous projection of 6.5 GW in September 2014). This brings the country's cumulative PV total to 18.3 GW. However, according to IHS, U.S. deployment amounted to 7 GW in 2014, or 800 MW higher than SEIA reported. In June 2014 Barclays downgraded bonds of U.S. utility companies. Barclays expects more competition by a growing self-consumption due to a combination of decentralized PV-systems and residential electricity storage. This could fundamentally change the utility's business model and transform the system over the next ten years, as prices for these systems are predicted to fall. For 2015, annual PV installations are predicted to increase to 8.1 GW (cumulative of 26.4 GW) by the end of the year. Other sources see U.S. deployment to increase by approximately 9 GW in 2015, before peaking in 2016. In May 2015, SEIA predicted U.S. PV-market to grow by 25% to 50% in 2016, with a cumulative PV capactiy of 40 GW by the end of 2016. Roughly, this translates into 13 GW of added capacity in 2016.

- Europe

By 2020, the European Photovoltaic Industry Association (EPIA) expects PV capacity to pass 150 GW. It finds the EC-supervised national action plans for renewables (NREAP) turned out to be too conservative, as the goal of 84 GW of solar PV by 2020 had already been surpassed in 2014 (prelim. figures account for close to 88 GW by the end of 2014). For 2030, EPIA originally predicted solar PV to reach between 330 and 500 GW, supplying 10 to 15 percent of Europe's electricity demand. However, recent reassessments are more pessimistic and point to a 7 to 11 percent share, if no major policy changes are undertaken.

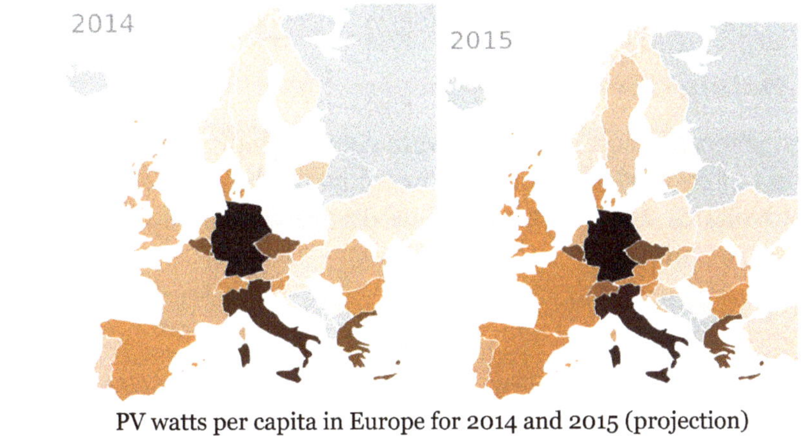

PV watts per capita in Europe for 2014 and 2015 (projection)

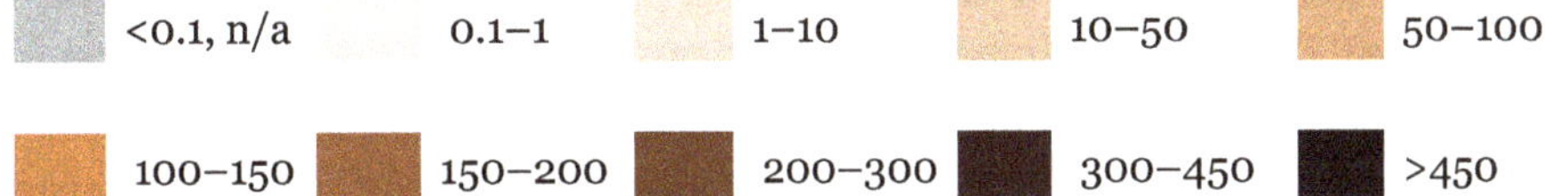

In 2014, overall European markets continued to decline despite strong growth in some countries. These countries, with their percentage-increase of total capacity, were, the United Kingdom (+80%), the Netherlands (+54%), Switzerland (+42%), Austria (+22%) and France (+20%), which rebounced significantly in terms of annual installations from 643 MW in 2013 to 927 MW in 2014. In most countries, however, deployment underperformed or, in some cases, growth of cumulative capacity even fell below the one percent mark. In descending percentage-order, cumulative installations only grew by 6.0% in Romania (69 MW), 5.2% in Germany (1,900 MW), 2.1% in Italy (385 MW), 2.1% in Belgium (65 MW), 0.4% in Spain (22 MW), 0.2% in Bulgaria (1.6 MW), 0.08% in Slovakia (0.4 MW), 0.06% in Greece (16 MW), and 0.01% in the Czech Republic (1.7 MW).

The United Kingdom had the strongest percentage growth and became the fourth largest PV installer worldwide after China, Japan and the United States. In 2014, the country installed more than 2.2 GW (vs. 1.1 GW in 2013) and cumulative installation increased by 80% to 5.1 GW by the end of the year. The booming utility-scale installations were partially explained by the upcoming closure of the appealing renewable obligation certificates (ROC) scheme in March 2015. This may also bring another record deployment in 2015, as IHS forecasts 3.5 GW of installations for the year.

In Germany and Italy, the rate of new installations continued to decline in 2014 and is expected to remain unchanged or even decline to 1.3 GW for 2015. In 2014, Germany installed 1,926 MW, down 36 percent from 3,300 MW deployed in 2013. During the period of 2010–2012, the country was the world's leader installing more than 7 GW annually. New cumulative capacity of 38.2 GW corresponds to 475 watts per inhabitant. Italy installed 385 MW, much less than previously expected and down from 1.5 GW deployed in 2013. Overall capacity of 18.5 GW translates into 304 watts per inhabitant. Solar PV now contributes significantly to domestic net-electricity consumption in Italy (7.9%), Greece (7.6%) and Germany (7.0%).

History of Leading Countries

Since the 1950s, when the first solar cells were commercially manufactured, there has been a succession of countries leading the world as the largest producer of electricity from solar photovoltaics. First it was the United States, then Japan, followed by Germany, and currently China.

United States (1954–1996)

The United States, inventor of modern solar PV, was the leader of installed capacity for many years. Based on preceding work by Swedish and German engineers, the American engineer Russell Ohl at Bell Labs patented the first modern solar cell in 1946. It was also there at Bell Labs where the first practical c-silicon cell was developed in 1954. Hoffman Electronics, the leading manufacturer of silicon solar cells in the 1950s and 1960s, improved on the cell's efficiency, produced solar radios, and equipped Vanguard I, the first solar powered satellite launched into orbit in 1958.

PV Capacity of Leading Countries (MW)

In 1977 US-President Jimmy Carter installed solar hot water panels on the White House promoting solar energy and the National Renewable Energy Laboratory, originally named *Solar Energy Research Institute* was established at Golden, Colorado. In the 1980s and early 1990s, most photovoltaic modules were used in stand-alone power systems or powered consumer products such as watches, calculators and toys, but from around 1995, industry efforts have focused increasingly on developing grid-connected rooftop PV systems and power stations. By 1996, solar PV capacity in the US amounted to 77 megawatts–more than any other country in the world at the time. Then, Japan stayed ahead.

Japan (1997–2004)

Japan took the lead as the world's largest producer of PV electricity, after the city of Kobe was hit by the Great Hanshin earthquake in 1995. Kobe experienced severe power outages in the aftermath of the earthquake, and PV systems were then considered as a temporary supplier of power during such events, as the disruption of the electric grid paralyzed the entire infrastructure, including gas stations that depended on electricity to pump gasoline. Moreover, in December of that same year, an accident occurred at the multibillion-dollar experimental Monju Nuclear Power Plant. A sodium leak caused a major fire and forced a shutdown (classified as INES 1). There was massive public outrage when it was revealed that the semigovernmental agency in charge of Monju had tried to cover up the extent of the accident and resulting damage. Japan remained world leader in photovoltaics until 2004, when its capacity amounted to 1,132 megawatts. Then, focus on PV deployment shifted to Europe.

Germany (2005–2014)

In 2005, Germany took the lead from Japan. With the introduction of the Renewable Energy Act in 2000, feed-in tariffs were adopted as a policy mechanism. This policy established that renewables have priority on the grid, and that a fixed price must be paid for the produced electricity over a 20-year period, providing a guaranteed return on investment irrespective of actual market prices. As a consequence, a high level of investment security lead to a soaring number of new photovoltaic installations that peaked in 2011, while investment costs in renewable technologies were brought down considerably. Germany's installed PV capacity is now approaching the 40,000 megawatt mark.

China (2015–present)

China surpassed Germany's capacity by the end of 2015, becoming the world's largest producer of photovoltaic power. China's rapid PV growth is expected to continue.

History of Market Development

Prices and Costs (1977–present)

The average price per watt has dropped drastically for solar cells over the last few decades. While in 1977 prices for crystalline silicon cells were about $77 per watt, average spot prices in June 2014 were as low as $0.36 per watt or 200 times less than almost forty years ago. Prices for thin-film solar cells and for c-Si solar panels were around $.60 per watt. In 2015, module and cell prices declined even further.

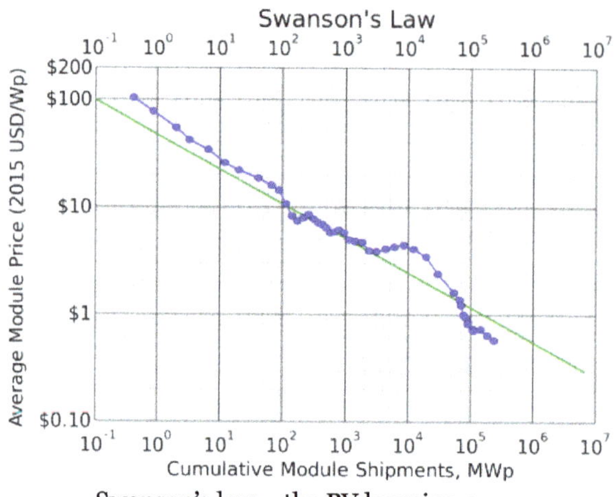

Swanson's law – the PV learning curve

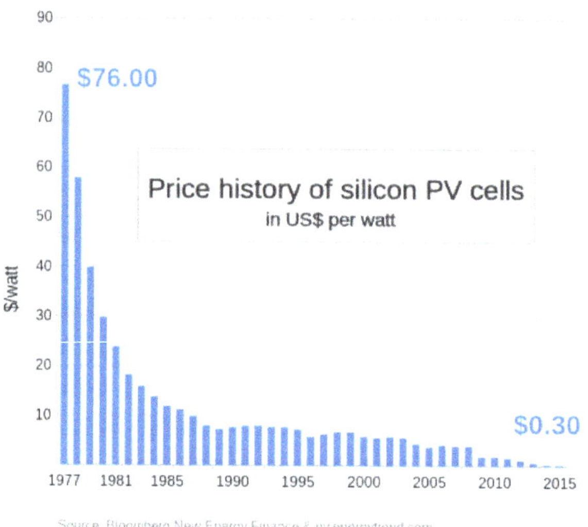

Price decline of c-Si solar cells

Type of cell or module	Price per watt
High efficiency multi-Si cell (>17.8%)	$0.325
Taiwanese multi-Si cell	$0.322
Chinese multi-Si cell	$0.294
Mono-Si cell	$0.335
Module (multi-Si)	$0.532
Module (mono-Si)	$0.595
Source: *EnergyTrend, price quotes, average prices, 2015*	

This price trend is seen as evidence supporting Swanson's law, an observation similar to the famous Moore's Law that states that the per-watt cost of solar cells and panels fall by 20 percent for every doubling of cumulative photovoltaic production. A 2015 study shows price/kWh dropping by 10% per year since 1980, and predicts that solar could contribute 20% of total electricity consumption by 2030.

In its 2014 edition of the *Technology Roadmap: Solar Photovoltaic Energy* report, the International Energy Agency (IEA) published prices for residential, commercial and utility-scale PV systems for eight major markets as of 2013 *(see table below)*. However, IEA's figures for the U.S seem to be controversial, as DOE's SunShot Initiative report states lower prices, although being published at the same time and referring to the same period. Prices have since fallen further. For 2014, the SunShot Initiative modeled U.S. system prices to be in the range of $1.80 to $3.29 per watt. Other sources identify similar price ranges of $1.70 to $3.50 for the different market segments in the U.S., and in the highly penetrated German market, prices for residential and small commercial rooftop systems of up to 100 kW declined to $1.36 per watt (€1.24/W) by the end of 2014. In 2015, Deutsche Bank estimated costs for small residential rooftop systems in the U.S. around $2.90 per watt. Costs for utility-scale systems in China and India were estimated as low as $1.00 per watt. As of mid-2015, a residential 5 kW-system in Australia costs on average about AU$1.60, or US$1.23 per watt.

Typical PV system prices in 2013 in selected countries (USD)

USD/W	Australia	China	France	Germany	Italy	Japan	United Kingdom	United States
Residential	1.8	1.5	4.1	2.4	2.8	4.2	2.8	4.9[1]
Commercial	1.7	1.4	2.7	1.8	1.9	3.6	2.4	4.5[1]
Utility-scale	2.0	1.4	2.2	1.4	1.5	2.9	1.9	3.3[1]

Source: *IEA – Technology Roadmap: Solar Photovoltaic Energy report, September 2014'* U.S figures are lower in DOE's Photovoltaic System Pricing Trends

Technologies (1990–Present)

With the advances in conventional crystalline silicon (c-Si) technology in recent years, and the falling cost of the polysilicon since 2009, that followed after a period of severe shortage of silicon feedstock, pressure increased on manufacturers of commercial thin-film PV technologies, including amorphous thin-film silicon (a-Si), cadmium telluride (CdTe), and copper indium gallium diselenide (CIGS), leading to the bankruptcy of several, once highly touted thin-film companies. The sector continues to face price competition from Chinese crystalline silicon cell and module manufacturers, and some companies together with their patents were sold below cost.

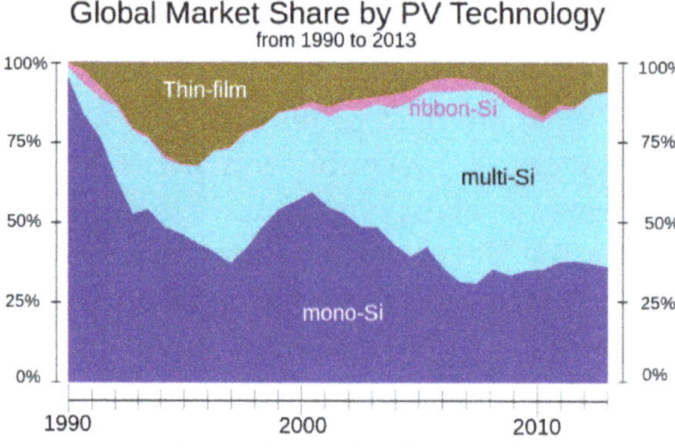

Market-share of PV technologies since 1990

- CIGS technology

Copper indium gallium selenide (CIGS) is the name of the semiconductor material the technology is based on. One of the largest producers of CIGS photovoltaics is the Japanese company Solar Frontier with a manufacturing capacity in the gigawatt-scale. The latest CIS line technology includes modules with conversion efficiencies of over 15%. The company profits from the booming Japanese market and attempts to widen its international business. However, several prominent manufacturers couldn't stand the pressure caused by advances in conventional crystalline silicon technology of recent years. The company Solyndra ceased all business activity and filed for Chapter 11 bankruptcy in 2011, and Nanosolar, also a CIGS manufacturer, closed its doors in 2013. Although both companies produced CIGS solar cells, it has been pointed out, that the failure was not due to the technology but rather because of the companies themselves, using a flawed architecture, such as, for example, Solyndra's cylindrical substrates.

- CdTe technology

The U.S.-company First Solar, a leading manufacturer of CdTe, has been building several of the world's largest solar power stations, such as the Desert Sunlight Solar Farm and Topaz Solar Farm, both in the Californian desert with a staggering 550 MW capacity each, as well as the 102 MW_{AC} Nyngan Solar Plant in Australia, the largest PV power station in the Southern Hemnisphere, commissioned in mid-2015. The company successfully produces CdTe-panels with a steadily increasing efficiency and declining cost per watt, as reported in 2013. CdTe has the lowest energy payback time of all mass-produced PV technologies, and can be as short as eight months in favorable locations. The company Abound Solar, also a manufacturer of cadmium telluride modules, went bankrupt in 2012.

- a-Si technology

In 2012, ECD solar, once one of the world's leading manufacturer of amorphous silicon (a-Si) technology, filed for bankruptcy in Michigan, United States. Swiss

OC Oerlikon divested its solar division that produced a-Si/μc-Si tandem cells to Tokyo Electron Limited. In 2014, the Japanese electronics and semiconductor company announced the closure of its micromorph technology development program. Other companies that left the amorphous silicon thin-film market include DuPont, BP, Flexcell, Inventux, Pramac, Schuco, Sencera, EPV Solar, NovaSolar (formerly OptiSolar) and Suntech Power that stopped manufacturing a-Si modules in 2010 to focus on crystalline silicon solar panels. In 2013, Suntech filed for bankruptcy in China.

Silicon Shortage (2005–2008)

In the early 2000s, prices for polysilicon, the raw material for conventional solar cells, were as low as $30 per kilogram and silicon manufacturers had initially no incentive to expand production by additional investments.

However, a severe silicon shortage came along in 2005, when governmental programmes sparked the deployment of solar PV to rise by 75% in Europe. In addition, the demand for silicon from semiconductor manufacturers was growing as well. Since the amount of silicon needed for semiconductors makes up a much smaller portion of production costs, manufacturers were able to outbid solar companies for the available silicon in the market.

Polysilicon prices since 2004. As of October 2015, the ASP for polysilicon stands at $15.30/kg

Initially, the incumbent polysilicon producers were slow to respond to rising demand for solar applications, because of their painful experiences with over-investment in the past. Silicon prices sharply rose to about $80 per kilogram, and reached as much as $400/kg for long-term contracts and spot prices. In 2007, the constraints on silicon became so severe that the solar industry was forced to idle about a quarter of its cell and module manufacturing capacity—an estimated 777 MW of the then available production capacity. The shortage also provided silicon specialists with both the cash and an incentive to develop new technologies and several new producers entered the market.

Early responses from the solar industry focused on improvements in the recycling of silicon. When this potential was exhausted, companies have been taking a harder look at alternatives to the conventional Siemens process.

As it takes about three years to build a new polysilicon plant, the shortage prolonged until 2008. Prices for conventional solar cells remained constant or even rose slightly during the period of silicon shortage from 2005 to 2008. This is notably seen as a "shoulder" that sticks out in the Swanson's PV-learning curve and it was feared that a prolonged shortage could delay solar power to become competitive with conventional energy prices without subsidies.

In the meantime the solar industry lowered the number of grams-per-watt by reducing wafer thickness and kerf loss, increased yields in all manufacturing steps, reducing module loss, and continuously raised panel efficiency. Finally, the ramp up of polysilicon production alleviated worldwide markets from the scarcity of silicon in 2009 and subsequently lead to an overcapacity with sharply declining prices in the photovoltaic industry for the following years.

Solar Overcapacity (2009–2013)

As the polysilicon industry had started to build additional large production capacities during the shortage period, prices dropped as low as $15 per kilogram forcing some producers to suspend production or exit the sector. Since then, prices for silicon have stabilized around $20 per kilogram and the booming solar PV market has also helped to reduce the enormous global overcapacity since 2009. However, overcapacity in the PV industry continues to persist. In 2013, global record deployment of 38 GW (updated EPIA figure) was still much lower than China's annual production capacity of approximately 60 GW. Continued overcapacity was further reduced by significantly lowering solar module prices and, as a consequence, many manufacturers could no longer cover costs or remain competitive. As worldwide growth of PV deployment continues and will likely break another record in 2014, the gap between overcapacity and global demand is expected to close in the next few years.

IEA-PVPS published historical data for the worldwide utilization of solar PV module production capacity that displays a slow return to normalization in manufacture in recent years. The utilization rate is the ratio of production capacities versus actual production output for a given year. A low of 49% was reached in 2007 and reflects the peak of the silicon shortage that idled a significant share of the module production capacity. As of 2013, the utilization rate had recovered somewhat and increased to 63%.

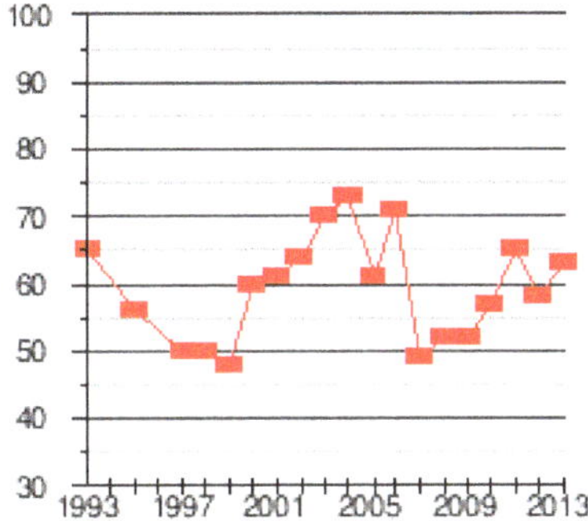

Utilization rate of solar PV module production capacity in % since 1993

Anti-dumping Duties (2012–Present)

After anti-dumping petition were filed and investigations carried out, the United States imposed tariffs of 31 percent to 250 percent on solar products imported from China in 2012. A year later, the EU also imposed definitive anti-dumping and anti-subsidy measures on imports of solar panels from China at an average of 47.7 percent for a two-year time span.

Shortly thereafter, China, in turn, levied duties on U.S. polysilicon imports, the feedstock for the production of solar cells. In January 2014, the Chinese Ministry of Commerce set its anti-dumping tariff on U.S. polysilicon producers, such as Hemlock Semiconductor Corporation to 57%, while other major polysilicon producing companies, such as German Wacker Chemie and Korean OCI were much less affected. All this has caused much controversy between proponents and opponents and is subject of current debate.

History of Deployment

Deployment figures on a global, regional and nationwide scale are well documented since the early 1990s. While worldwide photovoltaic capacity has been growing continuously, deployment figures by country are much more dynamic, as they depend strongly on national policies. A number of organizations release comprehensive reports on PV deployment on a yearly basis. They include annual and cumulative deployed PV capacity, typically given in watt-peak, a break-down by markets, as well as in-depth analysis and forecasts about future trends.

Worldwide Annual Deployment

Due to the exponential nature of PV deployment, about 83 percent of the overall capacity has been installed during the last five years from 2011 to 2015 *(see pie-chart; projected figure for 2015)*. Since the 1990s, and except for 2012, each year has been a record-breaking year in terms of newly installed PV capacity.

Worldwide Cumulative

Worldwide growth of solar PV capacity has been fitting an exponential curve since 1992. Tables below show global cumulative nominal capacity by the end of each year in megawatts, and the year-to-year increase in percent. In 2014, global capacity is expected to grow by 33 percent from 138,856 to 185,000 MW. This corresponds to an exponential growth rate of 29 percent or about 2.4 years for current worldwide PV capacity to double. Exponential growth rate: $P(t) = P_0 e^{rt}$, where P_o is 139 GW, growth-rate r 0.29 (results in doubling time t of 2.4 years).

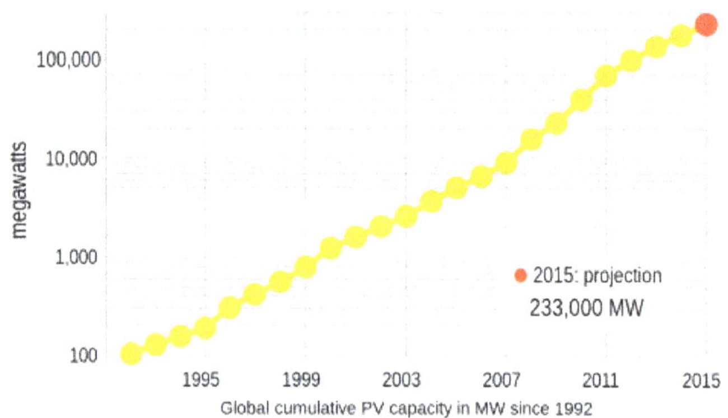

Worldwide cumulative PV capacity on a semi log chart since 1992

The following table contains data from four different sources. For 1992–1995: compiled figures of 16 main markets *(see section All time PV installations by country).* For 1996–1999: BP-Statistical Review of world energy (Historical Data Workbook) for 2000–2013: EPIA Global Outlook on Photovoltaics Report and for 2014, preliminary figures are based on IEA-PVPS' snapshot report

1990s			
Year	**Capacity[A] MW_p**	**Δ%[B]**	**Refs**
1991	*n.a.*	–	c
1992	105	n.a.	c
1993	130	24%	c
1994	158	22%	c
1995	192	22%	c
1996	309	61%	
1997	422	37%	
1998	566	34%	
1999	807	43%	
2000	1,250	55%	

2000s			
Year	**Capacity[A] MW$_p$**	**Δ%[B]**	**Refs**
2001	1,615	27%	
2002	2,069	28%	
2003	2,635	27%	
2004	3,723	41%	
2005	5,112	37%	
2006	6,660	30%	
2007	9,183	38%	
2008	15,844	73%	
2009	23,185	46%	
2010	40,336	74%	

2010s			
Year	**Capacity[A] MW$_p$**	**Δ%[B]**	**Refs**
2011	70,469	75%	
2012	100,504	43%	
2013	138,856	38%	
2014	178,391	28%	
2015	229,300	29%	
2016			
2017			
2018			
2019			
2020			

Legend:

^A Worldwide, cumulative nameplate capacity in megawatt-peak MW$_p$, (re-) calculated in DC power output.

^B annual increase of cumulative worldwide PV nameplate capacity in percent.

^C figures of 16 main markets, including Australia, Canada, Japan, Korea, Mexico, European countries, and the United States.

Deployment Reports

List of reports - Global Market Outlook for Photovoltaics

- 2014 – Global Market Outlook for Photovoltaics 2014–2018
- 2013 – Global Market Outlook for Photovoltaics 2013–2017

- 2012 – Global Market Outlook for Photovoltaics until 2016
- 2011 – Global Market Outlook for Photovoltaics until 2015
- 2010 – Global Market Outlook for Photovoltaics until 2014
- 2009 – Global Market Outlook for Photovoltaics until 2013
- 2008 – Global Market Outlook for Photovoltaics until 2012

Most PV deployment figures in this article are provided by the European Photovoltaic Industry Association in the "Global Outlook for Photovoltaics" report, the *Observatoire des énergies renouvelables* or EurObserv'ER's "Photovoltaic Barometer" report, and the IEA-PVPS (photovoltaic power systems) "Snapshot" and "Trends" report.

History of European PV deployment in *watts per capita* since 1992.

<0.1, n/a	50–100	300–450
0.1–1	100–150	>450
1–10	150–200	
10–50	200–300	

- EPIA

  The European Photovoltaic Industry Association (EPIA) represents members of the entire PV industry from silicon producers to cells and module manufactures and PV systems installers to PV electricity generation as well as marketing and sales. EPIA releases its annual *Global Market Outlook for Photovoltaics* report in May/June.

- PV-Barometer

List of reports - Photovoltaic Barometer report

- 2015 – figures for year 2013 and 2014
- 2014 – figures for year 2012 and 2013
- 2013 – figures for year 2011 and 2012
- 2012 – figures for year 2010 and 2011
- 2011 – figures for year 2009 and 2010
- 2010 – figures for year 2008 and 2009

EUROBSER'VER (*Observatoire des énergies renouvelables*) was set up in 1980, and is composed of engineers and experts releasing the *Photovoltaic Barometer* report containing early, year-end PV deployment figures for the 28 member states of the European Union. Eurobserver works closely together with several French ministries and is co-founded by the European Commission's IEE programm.

- IEA-PVPS

List of reports - PVPS Snapshot of Global PV and Trends

- 2015 – Snapshot of Global PV 1992-2014
- 2014 – Trends 2014 in Photovoltaic Applications – Survey report of selected IEA countries between 1992 and 2013
- 2014 – Snapshot of Global PV 1992-2013
- 2014 – Trends 2013 in Photovoltaic Applications – Survey report of selected IEA countries between 1992 and 2012
- 2013 – Snapshot of Global PV 1992-2012
- 2011 – Trends in Photovoltaic Applications – Survey report of selected IEA countries between 1992 and 2010
- 2010 – Trends in Photovoltaic Applications – Survey report of selected IEA countries between 1992 and 2009
- 2009 – Trends in Photovoltaic Applications – Survey report of selected IEA

countries between 1992 and 2008

- 2007 – Trends in Photovoltaic Applications – Survey report of selected IEA countries between 1992 and 2006

The IEA Photovoltaic Power Systems Programme (PVPS) is one of the collaborative R&D agreements established within the IEA and, since its establishment in 1993, the PVPS participants have been conducting a variety of joint projects in the application of photovoltaic conversion of solar energy into electricity. Its annual "Snapshot" report is released in early April and provides the first and detailed figures of worldwide PV-deployment of the previous year. An overview of all international statistics PDF reports since 1995 can be found on IEA-PVPS' Statistic Reports website.

Deployment by Country

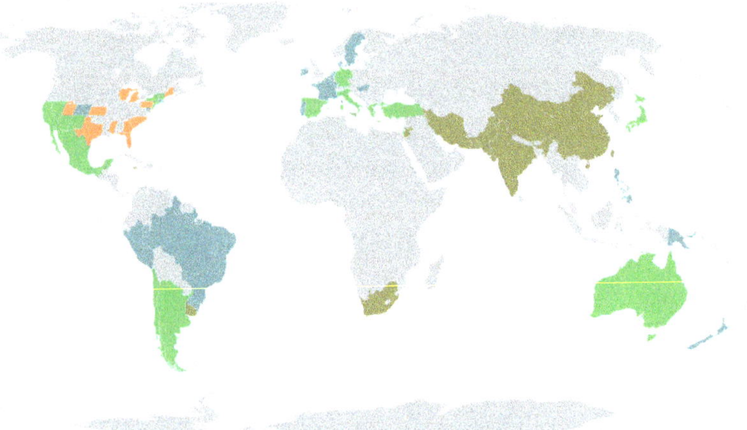

Grid parity for solar PV systems around the world

- Reached grid-parity before 2014
- Reached grid-parity after 2014
- Reached grid-parity only for peak prices
- U.S. states poised to reach grid-parity

2015				
2014				
2013				
2012				
2011				

2010				
2009				
2008				
2004				

2013				
2011				
2010				

2015

2014

2014 – Year-end PV Capacity by Country in MW

Global PV capacity by country in MW_p 2014[d]			
Country	**Capacity 2014**	**Added 2014**	**Source[a]**
Australia	4,136	910	*IEA-PVPS snapshot*
Austria	767	140	*EPIA-GMO*
Belgium	3,115	75	*IEA-PVPS AR2014*
Brazil	n.a.	22	
Bulgaria	1,022	2	*IEA-PVPS snapshot*
Canada[b,c]	1,710	500	*IEA-PVPS snapshot*
Chile	368	365	*IEA-PVPS snapshot*
China	28,050	10,640	*IEA-PVPS AR2014*
Croatia	33	13	*EPIA-GMO*
Cyprus	65	30	*PV Barometer*
Czech Republic	2,134	2	*IEA-PVPS snapshot*
Denmark	608	47	*EPIA-GMO*
Finland	10	–	*PV Barometer*
France	5,632	927	*EPIA-GMO*
Germany	38,235	1,898	*EPIA-GMO*
Greece	2,596	17	*EPIA-GMO*

Guatemala	n.a.	6	
Honduras	n.a.	5	
Hungary	38	3	*PV Barometer*
India	2,936	616	*IEA-PVPS snapshot*
Ireland	1	0.1	*PV Barometer*
Israel	731	250	*IEA-PVPS snapshot*
Italy	18,313	385	*EPIA-GMO*
Japan[(b)]	23,300	9,700	*IEA-PVPS snapshot*
Latvia	2	–	*PV Barometer*
Lithuania	68	–	*PV Barometer*
Luxembourg	110	15	*PV Barometer*
Malaysia	160	87	*IEA-PVPS snapshot*
Malta	54	26	*PV Barometer*
Mexico	176	64	*IEA-PVPS snapshot*
Netherlands[(c)]	1,042	400	*EPIA-GMO*
Norway	13	2	*IEA-PVPS snapshot*
Philippines	n.a.	n.a.	*not yet avail.*
Poland	34	27	*EPIA-GMO*
Portugal	414	115	*EPIA-GMO*
Romania	1,223	72	*EPIA-GMO*
Saudi Arabia	n.a.	n.a.	*not yet avail.*
Slovakia	524	0.4	*EPIA-GMO*
Slovenia	256	8	*PV Barometer*
South Africa	922	800	*IEA-PVPS snapshot*
South Korea[(c)]	2,384	909	*IEA-PVPS snapshot*
Spain[(b)]	5,388	22	*EPIA-GMO*
Sweden	79	36	*IEA-PVPS snapshot*
Switzerland	1,046	320	*EPIA-GMO*
Taiwan	776	400	*IEA-PVPS snapshot*
Thailand	1,299	475	*IEA-PVPS snapshot*
Turkey	58	40	*IEA-PVPS snapshot*

Ukraine	n.a.	n.a.	*not yet avail.*
United Kingdom	5,230	2,402	*EPIA-GMO*
United States	18,280	6,201	*IEA-PVPS snapshot*

[(a)] Data sources:

- IEA-PVPS, Snapshot of Global PV 1992–2014 report
- IEA-PVPS, Annual Report 2014 (AR2014)
- Photovoltaic Barometer, April 2015
- EPIA-SolarPower Europe, Global Market Outlook, June 2015

[(b)] Reconverted to watt-peak from MW_{AC}

[(c)] Official estimates only. Likely to change.

[(d)] Note: nameplate capacity always given in MW

Photovoltaic Barometer report – PV capacity in the European Union in 2014

Country	Added 2014 (MW)			Total 2014 (MW)				Generation 2014	
	off-grid	on-grid	Capacity	off-grid	on-grid	Capacity	Watt per-capita	in GWh	in%
Austria	–	140.0	140.0	4.5	766.0	770.5	90.6	766.0	–
Belgium	–	65.2	65.2	0.1	3,105.2	3,105.3	277.2	2,768.0	–
Bulgaria	–	1.3	1.3	0.7	1,019.7	1,020.4	140.8	1,244.5	–
Croatia	0.2	14.0	14.2	0.7	33.5	34.2	8.1	35.3	–
Cyprus	0.2	29.7	30.0	1.1	63.6	64.8	75.5	104.0	–
Czech Republic	–	–	–	0.4	2,060.6	2,061.0	196.1	2,121.7	–
Denmark	0.1	29.0	29.1	1.5	600.0	601.5	106.9	557.0	–
Estonia	–	–	–	0.1	–	0.2	0.1	0.6	–

Finland	–	–	–	10.0	0.2	10.2	1.9	5.9	–
France	0.1	974.9	975.0	10.8	5,589.0	4,697.6	87.6	5,500.0	–
Germany	–	1,899.0	1,899.0	65.0	38,236.0	38,301.0	474.1	34,930.0	–
Greece	–	16.9	16.9	7.0	2,595.8	2,602.8	236.8	3,856.0	–
Hungary	0.1	3.1	3.2	0.7	37.5	38.2	3.9	26.8	–
Ireland	0.0	0.0	0.1	0.9	0.2	1.1	0.2	0.7	–
Italy	1.0	384.0	385.0	13.0	18,437.0	18,450.0	303.5	23,299.0	–
Latvia	–	–	–	–	1.5	1.5	0.8	0.0	–
Lithuania	–	–	–	0.1	68.0	68.1	23.1	73.0	–
Luxembourg	–	15.0	15.0	–	110.0	110.0	200.1	120.0	–
Malta	–	26.0	26.0	–	54.2	54.2	127.5	57.8	–
Netherlands	–	361.0	361.0	5.0	1,095.0	1,100.0	65.4	800.0	–
Poland	0.5	19.7	20.2	2.9	21.5	24.4	0.6	19.2	–
Portugal	1.2	115.0	116.2	5.0	414.0	419.0	40.2	631.0	–
Romania	–	270.5	270.5	–	1,292.6	1,292.6	64.8	1,355.2	–
Slovakia	–	2.0	2.0	0.1	590.0	590.1	109.0	590.0	–
Slovenia	–	7.7	7.7	0.1	255.9	256.0	124.2	244.6	–
Spain	0.3	21.0	21.3	25.5	4,761.8	4,787.3	102.7	8,211.0	–
Sweden	1.1	35.1	36.2	9.5	69.9	79.4	8.2	71.5	–

United Kingdom	–	2,448.0	2,448.0	2.3	5,228.0	5,230.3	81.3	3,931.0	–
European Union	**4.9**	**6,878.4**	**6,883.3**	**167.1**	**86,506.8**	**86,673.9**	**171.5**	**91,319.7**	–
Country	**off-grid**	**on-grid**	**Capacity**	**off-grid**	**on-grid**	**Capacity**	**Watt per-capita**	**in GWh**	**in %**
	Added 2014 (MW)			**Total 2014 (MW)**				**Generation 2014**	

2013

2013 – Year-end PV capacity by country in MW

Worldwide

In 2013, worldwide deployment of solar PV amounted to almost 40 GW (39,953 MW)—an increase of about 35 percent over the previous year. Cumulated capacity increased by 38 percent to more than 139 GW. This is sufficient to generate at least 160 terawatt-hours (TWh) or 0.85 percent of the world's total electricity consumption of 18,400 TWh.

Worldwide installed capacity in 2013					
Report	**Cumulative (MW_p)**	**Installed (MW_p)**	**Release date**	**Type**	**Ref**
IEA-PVPS snapshot	>134,000	>36,900	April 2014	preliminary figures	
EPIA outlook	138,856	38,352	June 2014	detailed figures	
IEA-PVPS trends	139,795	39,953	October 2014	final figures	

An uncertainty in Chinese deployment was the reason for a significant discrepancy between EPIA's *outlook* and IEA's *Trends* reports. EPIA did not take 1.1 GW of additional Chinese deployment into account, since at the time it was unsure whether these installations were connected to the grid. In IEA's final *Trends* report, the confirmed figures not only raised domestic Chinese deployment from 11.8 GW to 12.92 GW, but also contributed to increase global figures to almost 40 GW.

Regions

In 2013, Asia has been the fastest growing region, with China and Japan accounting for 49% of worldwide deployment. About a quarter has been installed in Europe (10,975 MW). The remaining quarter of the 38,400 MW deployed in 2013 is split between North America and other countries.

Top 10 PV countries of Year 2013 in (MW)					
Total capacity			**Added capacity**		
1.	Germany	35,766	1.	China	12,920
2.	China	19,720	2.	Japan	6,968
3.	Italy	18,074	3.	United States	4,751
4.	Japan	13,599	4.	Germany	3,304
5.	United States	12,079	5.	Italy	1,620
6.	Spain	5,340	6.	UK	1,546
7.	France	4,733	7.	India	1,115
8.	UK	3,377	8.	Romania	1,100
9.	Australia	3,226	9.	Greece	1,043
10.	Belgium	3,009	10.	Australia	811

Data from the EPIA's *Global Market Outlook 2014-2018* report and partially updated with figures from the IEA-PVPS *Trends 2014* report

Europe is still the most developed region with a cumulative capacity of 81.5 GW, about 59 percent of the global total, followed by the Asia-Pacific region (APAC), including countries such as Japan, India and Australia with 22 GW or about 16 percent of worldwide cumulative capacity (due to its significance, China is excluded from the APAC region in all PV statistics and listed separately). European solar PV now covers 3 percent of the electricity demand and 6 percent of the peak electricity demand. However, deployment in Europe has slowed down by half compared to the record year of 2011, and will most likely continue to decrease. This is mainly due to the strong decline of new installations in Germany and Italy.

Cumulative capacity in the MEA (Middle East and Africa) region and ROW (rest of the world) accounted for less than 3 GW or about 2.2% of the global total. A great untapped potential remains for many of these countries, especially in the Sunbelt.

The top ten leading countries in terms of deployed and overall PV-capacity are shown above. Other mentionable PV deployments above the 100-megawatt mark included France (643 MW), Canada (445 MW), South Korea (445 MW), Thailand (437 MW), The Netherlands (360 MW), Switzerland (319 MW), Ukraine (290 MW), Austria (263 MW), Israel (244 MW), Belgium (237 MW), Taiwan (170 MW) Denmark (156 MW) and Spain (116 MW).

2013—Global PV capacity by country in MW_p			
Country	**Capacity 2013**	**Added 2013**	**Source**
Australia	3,226	811	*IEA-PVPS trends*

Austria	626	263	*IEA-PVPS trends*
Belgium	3,009	237	*IEA-PVPS trends*
Brazil	32	?	?
Bulgaria	1,020	10	*EPIA outlook*
Canada	1,210	445	*IEA-PVPS trends*
Chile	3	2	*original. Revised cume in 2015*
China	19,720	12,920	*IEA-PVPS trends*
Croatia	20	20	*EPIA outlook*
Cyprus	32	15	*EPIA outlook*
Czech Republic	2,175	88	*EPIA outlook*
Denmark	563	156	*IEA-PVPS trends*
Finland	11	0	*EPIA outlook*
France	4,733	643	*IEA-PVPS trends*
Germany	35,766	3,304	*IEA-PVPS trends*
Greece	2,579	1,043	*EPIA outlook*
Hungary	22	10	*EPIA outlook*
India	2,320	1,115	*IEA-PVPS trends*
Ireland	3	0	*EPIA outlook*
Israel	481	244	*IEA-PVPS trends*
Italy	18,074	1,620	*IEA-PVPS trends*
Japan	13,599	6,968	*IEA-PVPS trends*
Latvia	1	0	*EPIA outlook*
Lithuania	6	0	*EPIA outlook*
Luxembourg	30	0	*EPIA outlook*
Malaysia	73	48	*IEA-PVPS trends*
Malta	23	7	*EPIA outlook*
Mexico	112	60	*IEA-PVPS Trends*
Netherlands	723	360	*IEA-PVPS trends*
Norway	11	1	*IEA-PVPS trends*
Poland	7	1	*EPIA outlook*

Portugal	281	53	*IEA-PVPS trends*
Romania	1,151	1,100	*EPIA outlook*
Slovakia	524	1	*EPIA outlook*
Slovenia	212	11	*EPIA outlook*
South Africa	122	92	*IEA-PVPS revised in 2015*
South Korea	1,475	445	*IEA-PVPS trends*
Spain	5,340	116	*EPIA outlook*
Sweden	43	19	*IEA-PVPS trends*
Switzerland	756	319	*IEA-PVPS trends*
Taiwan	376	170	*IEA-PVPS trends*
Thailand	824	437	*IEA-PVPS trends*
Turkey	18	6	*IEA-PVPS trends*
Ukraine	616	290	*EPIA outlook*
United Kingdom	3,377	1,546	*IEA-PVPS trends*
United States	12,079	4,751	*IEA-PVPS trends*

Listed countries account for 39,666 MW or 99.3% of worldwide total installations of 39,953 MW in 2013.

(A) **Data sources** (Nameplate capacity in DC-output not AC):

- IEA-PVPS – Trends 2014 in Photovoltaic Applications
- EPIA – Global Market Outlook for Photovoltaics 2014–2018
- Find additional or alternative information in column *source* of corresponding record

In 2013, Europe added 11 gigawatts of new PV installation (including non-EU countries). It is still the most developed region with a cumulated total of 81.5 GW, about 59 percent of the worldwide installed capacity. Solar PV now covers 3 percent of the electricity demand and 6 percent of the peak electricity demand. However, European PV deployment has slowed down by half compared to the record year of 2011, and will most likely continue to decrease. This is mainly due to the strong decline of new installations in Germany and Italy.

Photovoltaic Barometer report – PV capacity in the European Union in 2013										
Country	**Added 2013 (MW)**			**Total 2013 (MW)**				**Generation 2013**	
	off-grid	**on-grid**	**Capacity**	**off-grid**	**on-grid**	**Capacity**	**Watt per-capita**	**in GWh**	**in %**

Austria	–	268.7	268.7	4.5	685.9	690.4	81.7	686.0	–
Belgium	–	214.9	215.0	0.1	2,983.3	2,983.4	267.3	2,352.0	–
Bulgaria	–	104.4	104.4	0.7	1,018.5	1,019.2	139.9	1,348.5	–
Croatia	–	17.2	17.2	0.5	21.2	21.7	5.1	12.3	–
Cyprus	0.1	17.5	17.5	0.9	33.9	34.8	40.2	45.8	–
Czech Republic	–	110.4	110.4	0.4	2,132.4	2,132.8	202.8	2,070.0	–
Denmark	0.2	155.0	155.2	1.4	530.0	531.4	94.8	490.0	–
Estonia	–	–	–	0.1	–	0.2	0.1	0.6	–
Finland	–	–	–	11.0	0.2	11.2	2.1	5.4	–
France	–	613.0	613.0	24.6	4,673.0	4,697.6	71.6	4,900.0	–
Germany	5.0	3,305.0	3,310.0	65.0	35,948	36,013.0	447.2	30,000.0	–
Greece	–	1,042.5	1,042.5	7.0	2,578.8	2,585.8	233.7	3,648.0	–
Hungary	0.1	3.0	3.1	0.6	14.8	15.4	1.6	9.3	–
Ireland	0.1	–	0.1	0.9	0.2	1.0	0.2	0.7	–
Italy	1.0	1,461.0	1,462.0	12.0	17,602.0	17,614.0	295.1	22,146.0	–
Latvia	–	–	–	–	1.5	1.5	0.7	n.a.	–
Lithuania	–	61.9	61.9	0.1	68.0	68.1	22.9	45.0	–
Luxembourg	–	23.3	23.3	–	100.0	100.0	186.2	50.0	–
Malta	–	6.0	6.0	–	24.7	24.7	58.7	30.1	–
Netherlands	–	300.0	300.0	5.0	660.0	665.0	39.6	582.0	–
Poland	0.2	0.4	0.6	2.4	1.8	4.2	0.1	4.0	–
Portugal	0.5	52.2	52.7	3.8	277.2	281.0	26.8	446.0	–
Romania	–	972.7	972.7	–	1,022.0	1,022.0	51.1	397.8	–

Country	Added 2013 (MW)			Total 2013 (MW)				Generation 2013	
	off-grid	on-grid	Capacity	off-grid	on-grid	Capacity	Watt per-capita	in GWh	in %
Slovakia	–	–	–	0.1	537.0	537.1	99.3	600.0	–
Slovenia	–	33.3	33.3	0.1	254.7	254.8	123.8	240.0	–
Spain	0.4	102.0	102.4	25.0	4,680.5	4,705.5	100.7	8,289.0	–
Sweden	1.1	17.9	19.0	8.4	34.7	43.1	4.5	38.8	–
United Kingdom	–	1,031.0	1,031.0	2.3	2,737.0	2,739.3	42.9	1,800.0	–
European Union	**8.7**	**9,913.5**	**9,922.2**	**177.0**	**78,621.2**	**78,798.2**	**155.8**	**80,236.0**	**–**

2012

2012 – Year-end PV capacity by country in MW

2012 – Year-end PV capacity by country in MW									
Country	**Capacity**							**Generation**	
	off-grid Δ	**on-grid Δ**	**Added**	**off-gri Σ**	**on-grid Σ**	**Total**	**Watt/Capita**	**GWh**	**%**
Germany	0.0	7,604	7,604	55.0	32,643	32,698	398	28,000	5.62%
Italy	1.0	3,577	3,578	11.0	16,350	16,361	273	18,800	6.70%
China	–	–	5,000	–	–	8,300	6	6,678	0.14%
United States	–	–	3,346	–	–	7,777	24	9,750	0.25%
Japan	–	–	2,000	–	–	7,000	55	6,600	0.77%
Spain	1.3	193	194	7.3	4,492	4,517	110	8,169	2.84%
France	0.0	1,079	1,079	24.6	4,003	4,028	61	4,000	0.91%
Belgium	0.0	599	599	0.1	2,650	2,650	241	2,115	2.90%
Australia	–	–	1,000	–	–	2,400	105	2,800	1.23%
Czech Republic	0	109	109	0.4	2,022	2,022	196	2,173	3.11%
United Kingdom	0	929	929	0	1,829	1,829	29	1,327	0.50%
Greece	0	912	912	7.0	1,536	1,543	–	1,239	4.26%

India	–	–	980	–	–	1,205	1	2,115	0.33%
Korea	–	–	252	–	–	1,064	22	920	0.20%
Bulgaria	0.0	721	721	0.7	933	933	123	534	3.40%
Canada	–	–	268	–	–	765	22	860	0.17%
Slovakia	0.0	30	30	0.1	517	517	95	500	1.99%
Austria	0.0	235	235	4.5	417	422	50	300	0.62%
Switzerland	–	–	200	–	–	410	53	370	0.64%
Denmark	0.0	375	375	1.7	390	392	70	114	1.09%
Ukraine	–	–	188	–	–	373	8	410	0.31%
Thailand	–	–	210	–	–	359	5	530	0.40%
Netherlands	0.0	175	175	2.0	316	321	16	200	0.21%
Israel	–	–	60	–	–	250	31	310	0.68%
Portugal	0.1	68	68	0.6	226	229	22	360	0.74%
Slovenia	0.0	117	117	24.6	217	217	97	121	1.70%
Taiwan	–	–	104	–	–	206	9	–	–
Mexico	–	–	15	–	–	38	0.3	83	0.04%
Luxembourg	0	7	7	0	47	47	59	30	0.44%
South Africa	–	–	40	–	–	41	0.08	–	–
Malaysia	–	–	22	–	–	36	1	34	0.04%
Romania	–	–	26	–	–	30	2	–	–
Sweden	0.8	7	8	2.3	17	24	2	21	0.01%
Malta	0	12	12	5.0	19	19	29	14	0.89%
Cyprus	0	7	7	0.8	16	17	11	20	0.31%
Brazil	–	–	12	–	–	17	0.1	–	–
Peru	–	–	15	–	–	15	0.5	–	–
Finland	0	0	0	11.0	0	11	–	8	0.00%
Norway	–	–	0	–	–	9	–	7	0.01%
Turkey	–	–	2	–	–	9	0.1	10	0.01%
Lithuania	0	6	6	0.0	6	6.1	2	2	0.05%
Chile	–	–	2	–	–	6	0.3	2	–
Hungary	0	1	1	0.5	3	3.74	–	5	0.01%

Country	off grid Δ	on grid Δ	Added	off grid Σ	on grid Σ	Total	Watt/ Capita	GWh	%
Poland	1.1	0	1	3.3	1	3.4	0.1	4	0.00%
Latvia	0	0	0	0.1	1.5	1.51	0.3	0	0.01%
Ireland	0	0	0	0.6	0.1	0.69	–	0	0.01%
Croatia	–	–	0.2	–	–	0.2	0.04	–	–
Estonia	0	0	0	0.13	0.02	0.15	0.1	0	0.00%
	Capacity							Generation	

2011

2011 – Year-end PV capacity by country in MW

2010

2010 – Year-end PV capacity by country in MW

2010 – Year-end PV capacity by country in MW									
Country	**off-grid Δ**	**on-grid Δ**	**Installed 2010**	**off grid Σ**	**on grid Σ**	**Total 2010**	**Watt/ Capita**	**Module Price US$/ Wp**	**Feed-in Tariff US$/ kW·h**
Germany	5	7,406	7,411	50	17,320	17,370	212.3	–	–
Spain	1	369	370	21.1	3,787	3,808	82.8	–	–
Japan	4.2	986.8	991.0	98.8	3,519	3,618	28.3	–	–
Italy	0.1	2,321	2,321	13.5	3,465	3,478	57.6	–	–
United States	31	887	918	440	2,094	2,534	8.1	1.48-2.36	–
Czech Republic	0	1,490	1,490	0.4	1,952.7	1,953	185.9	–	–
France	0.1	719.0	719.1	29.4	1,025	1,054	16.3	–	–
China	–	–	520.0	–	–	893.0	–	–	–
Belgium	0	213.4	213.4	0.1	787.4	787.5	72.6	–	–
South Korea	0	131.2	131.2	6.0	649.6	655.6	13.4	–	–
Australia	3.8	379.5	383.3	87.8	483.1	570.9	25.2	–	–
Canada	24.9	171.7	196.6	60.1	231.0	291.1	8.4	–	–
Greece	0.1	150.3	150.4	6.9	198.5	205.4	18.2	–	–
India	–	–	69.0	–	–	189.0	–	–	–

Country	off grid Δ	on grid Δ	In-stalled 2010	off grid Σ	on grid Σ	Total 2010	Watt/ Cap-ita	Module Price US$/ Wp	Feed-in Tariff US$/ kW·h
Switzerland	0.2	25.5	25.7	4	69.6	73.6	9.7	3.0-3.5	14.7
Netherlands	0.091	10.58	10.67	5	62.5	67.5	4.1	–	–
Austria	0.25	19.96	20.21	3.61	48.99	52.60	6.4	3.0-3.2	26.4
United Kingdom	0.155	6.922	7.077	0	26.4	26.4	0.4	–	–
Mexico	2.47	0.80	3.27	23.72	1.30	25.02	0.2	4-5	<36.2
Israel	0.5	21	21.5	2.9	21.63	24.53	3.4	3.6-5.1	12.5
Portugal	0.2	14.25	14.45	2.841	15.03	17.87	1.7	2.1-4.2	10.3-19.2
Malaysia	2	0.287	2.287	10	1.063	11.06	0.4	3.71	<13.5
Slovenia	0	6.9	6.9	0.1	8.9	9.0	4.1	–	–
Sweden	0.338	0.516	0.854	5.169	3.595	8.764	1.0	2.4-6.5	18.3-20.9
Norway	0.32	0	0.32	8.530	0.132	8.662	1.9	–	11.1–14.3
Finland	2.0	0	2.0	7.5	0.2	7.6	1.4	–	–
Luxembourg	0	1.8	1.8	0	5.7	5.7	52.4	–	–
Bulgaria	0	4.3	4.3	0	5.7	5.7	0.8	–	–
Denmark	0.2	1.2	1.3	0.540	4.025	4.565	0.8	2.8-4.7	37.5
Turkey	0.9	0.1	1	4.5	0.5	5	0.1	2.8-4.2	13.3

2009

2009 – Year-end PV capacity by country in MW

2009 – Year-end PV capacity by country in MW									
Country	off-grid Δ	on-grid Δ	In-stalled 2009	off-grid Σ	on-grid Σ	Total 2009	Watt/ Capi-ta	Mod-ule-Price US$/ Wp	Feed-in Tariff US$/ kW·h
Germany	5	3,840	3,845	45	9,800	9,845	119.6	2.1-3.5	31.5
Spain	0	60	60	31	3,492	3,523	76.1	1.6-3.5	–
Japan	3.8	479.2	483.0	95	2,533	2,627	20.7	4.3	19.1-25.8
United States	40	433.1	473.1	410	1,232	1,642	5.3	1.85-2.2	10.4
Italy	0	723	723	13	1168	1,181	20.3	2.2-3.0	23.6

Country	off-grid Δ	on-grid Δ	In-stalled 2009	off-grid Σ	on-grid Σ	Total 2009	Watt/ Capi-ta	Mod-ule-Price US$/ Wp	Feed-in Tariff US$/ kW·h
South Korea	0	84.4	84.4	5.9	436.0	441.9	9.1	1.9-2.0	13.3-19.6
France	0.2	250	250.2	23	407	430	6.7	1.5-2.8	–
Australia	10.56	68.57	79.13	83.91	99.7	183.65	8.3	2.3-4.7	10.2-15.6
Portugal	0.1	34.15	34.25	3.05	99.15	102.2	9.5	2.1-4.2	10.3-19.2
Canada	7.71	54.14	61.85	35.2	59.37	94.57	2.8	2.9	6.1
Switzerland	0.2	25.5	25.7	4	69.6	73.6	9.7	3.0-3.5	14.7
Netherlands	0.091	10.58	10.67	5	62.5	67.5	4.1	–	–
Austria	0.25	19.96	20.21	3.61	48.99	52.60	6.4	3.0-3.2	26.4
United Kingdom	0.155	6.922	7.077	1.75	27.85	29.59	0.4	–	–
Mexico	2.47	0.80	3.27	23.72	1.30	25.02	0.2	4-5	<36.2
Israel	0.5	21	21.5	2.9	21.63	24.53	3.4	3.6-5.1	12.5
Malaysia	2	0.287	2.287	10	1.063	11.06	0.4	3.71	<13.5
Sweden	0.338	0.516	0.854	5.169	3.595	8.764	1.0	2.4-6.5	18.3-20.9
Norway	0.32	0	0.32	8.530	0.132	8.662	1.9	–	11.1–14.3
Turkey	0.9	0.1	1	4.5	0.5	5	0.1	2.8-4.2	18.0
Denmark	0.2	1.2	1.3	0.540	4.025	4.565	0.8	2.8-4.7	37.5

2008

2008 – Year-end PV capacity by country in MW

2008 – Year-end PV capacity by country in MW							
Country	**off-grid Δ**	**on grid Δ**	**Installed 2008**	**off-grid Σ**	**on-grid Σ**	**Total 2008**	**Watt/ Capita**
Germany	4.5	1,500	1,504.5	40	5,300	5,340	64.7
Spain	1	60	2,661	31	3,323	3,354	77.1
Japan	0.7	224.6	225.3	90.8	2,053	2,144	16.8
United States	45	293	338	370	798.5	1,168.5	3.9

Italy	0.2	337.9	338.1	13.3	445	458.3	7.8
South Korea	0	276.3	276.3	5.9	351.6	357.5	7.3
France	0.4	104.1	104.5	22.9	156.8	179.7	2.9
Australia	6.9	15.1	22.0	73.3	31.2	104.5	5.1
Portugal	0.1	49.98	50.08	2.941	65.01	67.95	6.7
Netherlands	0.21	4.2	4.4	5.2	52	57.2	3.5
Switzerland	0.2	11.5	11.7	3.8	44	47.9	6.4
Canada	4.62	2.33	6.94	27.48	5.24	32.7	1.0
Austria	0.13	4.55	4.69	3.36	29.03	32.39	4.0
United Kingdom	0.12	4.3	4.42	1.59	20.92	22.51	0.3
Mexico	0.80	0.20	1.00	21.25	0.50	21.75	0.2
Malaysia	1.6	0.135	1.76	8	0.776	8.776	0.4
Norway	0.35	0	0.35	8.210	0.132	8.342	1.8
Sweden	0.275	10.403	1.678	4.83	3.08	7.91	0.9
Turkey	0.675	0.075	0.75	3.75	0.25	4	0.06
Denmark	0.55	0.135	0.190	0.440	2.825	3.265	0.8
Israel	0.6	0.6	1.21	2.4	0.62	3.03	0.4
Country	**off-grid Δ**	**on grid Δ**	**Installed 2009**	**off-grid Σ**	**on-grid Σ**	**Total 2009**	**Watt/capita**

2007

2007 – Year-end PV capacity by country in MW

Off grid refers to photovoltaics which are not grid connected. On grid means connected to the local electricity grid. Δ means the amount installed during the previous year. Σ means the total amount installed. Wp/capita refers to the ratio of total installed capacity divided by total population, or total installed Wp per person. Module price is average installed price, in Euros. kW·h/kWp·yr indicates the range of insolation to be expected. While National Report(s) may be cited as source(s) within an International Report, any contradictions in data are resolved by using only the most recent report's data. Exchange rates represent the 2006 annual average of daily rates (OECD Main Economic Indicators June 2007). Module Price: Lowest:2.5 EUR/Wp (2.83 USD/Wp) in Germany 2003. Uncited insolation data is from maps dating 1991–1995.

2007 – Year-end PV capacity by country in MW									
Country	**off-grid Δ**	**on-grid Δ**	**In-stalled 2007**	**off-grid Σ**	**on-grid Σ**	**To-tal 2007**	**Watt/ Capi-ta**	**Mod-ule-Price €/Wp**	**Feed-in Tar-iff EU¢/kWh**
Germany	35	1,100	1,135	35	3,827	3,862	46.8	4.0–5.3	51.8–56.8
Japan	1.562	208.8	210.4	90.15	1,829	1,919	15	2.96	Ended(2005)
United States	55	151.5	206.5	325	505.5	830.5	2.8	2.98	1.2–31.04(CA)
Spain	22	490	512	29.8	625.2	655	15.1	3.0–4.5	18.38–44.04
Italy	0.3	69.9	70.2	13.1	107.1	120.2	2.1	3.2–3.6	36.0–49.0
Australia	5.91	6.28	12.19	66.45	16.04	82.49	4.1	4.5–5.4	0–26.4(SA'08)
South Korea	0	42.87	42.87	5.943	71.66	77.60	1.6	3.50–3.84	56.5–59.3
France	0.993	30.31	31.30	22.55	52.68	75.23	1.2	3.2–5.1	30.0–55.0
Netherlands	0.582	1.023	1.605	5.3	48	53.3	3.3	3.3–4.5	1.21–9.7
Switzerland	0.2	6.3	6.5	3.6	32.6	36.2	4.9	3.18–3.30	9.53–50.8
Austria	0.055	2.061	2.116	3.224	24.48	27.70	3.4	3.6–4.3	>0
Canada	3.888	1.403	5.291	22.86	2.911	25.78	0.8	3.76	0–29.48(ON)
Mexico	0.869	0.15	1.019	20.45	0.3	20.75	0.2	5.44–6.42	None
United Kingdom	0.16	3.65	3.81	1.47	16.62	18.09	0.3	3.67–5.72	0–11.74(exprt)
Portugal	0.2	14.25	14.45	2.841	15.03	17.87	1.7	–	–
Norway	0.32	0.004	0.324	7.86	0.132	7.992	1.7	11.2	None
Sweden	0.271	1.121	1.392	4.566	1.676	6.242	0.7	3.24–7.02	None
Denmark	0.05	0.125	0.175	0.385	2.69	3.075	0.6	5.36–8.04	None
Israel	0.5	0	0.5	1.794	0.025	1.819	0.3	4.3	13.13–16.40
Country	**off-grid Δ**	**on-grid Δ**	**In-stalled 2007**	**off-grid Σ**	**on-grid Σ**	**To-tal 2007**	**Watt/ Capi-ta**	**Module Price €/Wp**	**Feed-in Tar-iff EU¢/kW·h**

2006

2006 – Year-end PV capacity by country in MW

Notes: While National Report(s) may be cited as source(s) within an International Report, any contradictions in data are resolved by using only the most recent report's data. Exchange rates represent the 2006 annual average of daily rates (OECD Main Economic Indicators June 2007) Module Price: Lowest:2.5 EUR/Wp (2.83 USD/Wp) in Germany 2003. Uncited insolation data is lifted from maps dating 1991–1995.

2005

2005 – Year-end PV capacity by country in MW

Original source gives these individual numbers and totals them to 37,500 kW. The 2004 reported total was 30,700 kW. With new installations of 6,800 kW, this would give the reported 37,500 kW.

2004

2004 – Year-end PV capacity by country in MW

Photovoltaic Thermal Hybrid Solar Collector

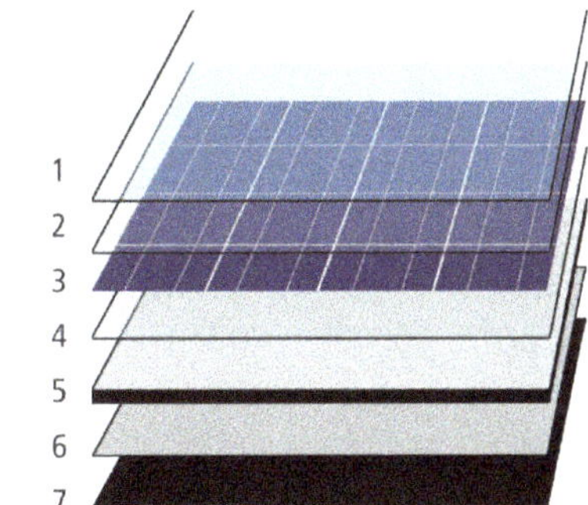

Schematic of a hybrid (PVT) solar collector:

1 - Anti-reflective glass

2 - EVA-encapsulant

3 - Solar PV cells

4 - EVA-encapsulant

5 - Backsheet (PVF)

6 - Heat exchanger (copper)

7 - Insulation (polyurethane)

Photovoltaic thermal hybrid solar collectors, sometimes known as hybrid PV/T systems or PVT, are systems that convert solar radiation into thermal and electrical energy. These systems combine a solar cell, which converts sunlight into electricity, with a solar thermal collector, which captures the remaining energy and removes waste heat from the PV module. The capture of both electricity and heat allow these devices to have higher exergy and thus be more overall energy efficient than solar photovoltaic (PV) or solar thermal alone. A significant amount of research has gone into developing PVT technology since the 1970s.

Photovoltaic cells suffer from a drop in efficiency with the rise in temperature due to increased resistance. Such systems can be engineered to carry heat away from the PV cells thereby cooling the cells and thus improving their efficiency by lowering resistance. Although this is an effective method, it causes the thermal component to under-perform compared to a solar thermal collector. Recent research showed that photovoltaic materials with low temperature coefficients such as amorphous silicon (a-Si:H) PV allow the PVT to be operated at high temperatures, creating a more symbiotic PVT system. This advantage can be tuned by controlling the dispatch strategy of thermal annealing cycles in any region of the world.

System Types

A number of PV/T collectors in different categories are commercially available and can be divided into the following categories:

- PV/T liquid collector
- PV/T air collector
- PV/Ta Liquid and air collector
- PV/T concentrator (CPVT)

PV/T Liquid Collector

The basic water-cooled design uses conductive-metal piping or plates attached to the back of a PV module. The fluid flow arrangement through the cooling element will determine which systems the panels are most suited to.

In a standard fluid based system, a working fluid, typically water, glycol or mineral oil is then piped through these pipes or plate chillers. The heat from the PV cells is conducted through the metal and absorbed by the working fluid (presuming that the working fluid is cooler than the operating temperature of the cells). In closed-loop systems this heat is either exhausted (to cool it), or transferred at a heat exchanger, where it flows to its application. In open-loop systems, this heat is used, or exhausted before the fluid returns to the PV cells. It is also possible to disperse nanoparticles in the liquid to create a liquid filter for PV/T applications. The basic advantage of this type of split configuration is that the thermal collector and the photovoltaic collector can operate at different temperatures.

PV/T Concentrator (CPVT)

A concentrator system has the advantage to reduce the amount of photovoltaic (PV) cells needed, such that somewhat more expensive and efficient multi-junction photovoltaic cells can be used that will maximize the ratio of produced high-value electrical power versus lower-value thermal power. A major limitation of high-concentrator (i.e. HCPV and HCPVT) systems is that they maintain their advantage over conventional c-Si/mc-Si collectors only in regions that remain consistently free of atmospheric aerosol contaminants (e.g. light clouds, smog, etc.). Concentrator system performance is especially degraded because 1) radiation is reflected and scattered outside of the small (often less than 1°-2°) acceptance angle of the collection optics, and 2) absorption of specific components of the solar spectrum causes one or more series junctions within the MJ cells to underperfom.

Concentrator systems also require reliable control systems to accurately track the sun and to protect the PV cells from damaging over-temperature conditions. Under ideal conditions, about 75% of the suns power directly incident upon such systems can be gathered as electricity and heat. For more details, see the discussion of CPVT within the article for concentrated photovoltaics.

References

- Palz, Wolfgang (2013). Solar Power for the World: What You Wanted to Know about Photovoltaics. CRC Press. pp. 131–. ISBN 978-981-4411-87-5.
- Luque, Antonio & Hegedus, Steven (2003). Handbook of Photovoltaic Science and Engineering. John Wiley and Sons. ISBN 0-471-49196-9.
- Gevorkian, Peter (2007). Sustainable energy systems engineering: the complete green building design resource. McGraw Hill Professional. ISBN 978-0-07-147359-0.
- Williams, Neville (2005). Chasing the Sun: Solar Adventures Around the World. New Society Publishers. p. 84. ISBN 9781550923124.
- Pearce, J.; Lau, A. (2002). "Net Energy Analysis for Sustainable Energy Production from Silicon Based Solar Cells". Solar Energy (PDF). p. 181. doi:10.1115/SED2002-1051. ISBN 0-7918-1689-3.
- Wolfe, Philip (2012). Solar Photovoltaic Projects in the Mainstream Power Market. Oxford: Routledge. p. 240. ISBN 978-0-415-52048-5.
- Eiffert, Patrina; Kiss, Gregory J. (2000). Building-Integrated Photovoltaic Designs for Commercial and Institutional Structures: A Source Book for Architect. p. 59. ISBN 978-1-4289-1804-7.
- Igor Bazovsky, Chapter 18: Reliability Design Considerations. In: Reliability Theory and Practice, 1963 (reprinted 2004), Pages 176-185, ISBN 978-0486438672
- Solar PV Module Costs to Fall to 36 Cents per Watt by 2017. Greentechmedia.com (2013-06-18). Retrieved on 2015-04-15.
- Harshavardhan Dinesh, Joshua M. Pearce, The potential of agrivoltaic systems, Renewable and Sustainable Energy Reviews, 54, 299-308 (2016).
- "World's Largest Hydro/PV Hybrid Project Synchronized". Corporate News. China State Power

Investment Corporation. 14 December 2014. Retrieved 22 July 2016.

- Canellas (et al), Claude (1 December 2015). "New French solar farm, Europe's biggest, cheaper than new nuclear". Reuters. Retrieved March 2016.
- "Enel Starts Production at its Largest Solar PV Project in Chile". Renewable Energy World. 31 May 2016. Retrieved 22 July 2016.
- S Hill, Joshua (January 22, 2016). "China Overtakes Germany To Become World's Leading Solar PV Country". Clean Technica. Retrieved August 16, 2016.
- "Oerlikon Divests Its Solar Business and the Fate of Amorphous Silicon PV". greentechmedia.com. Retrieved 27 January 2016.
- "NovaSolar, Formerly OptiSolar, Leaving Smoking Crater in Fremont". greentechmedia.com. Retrieved 27 January 2016.

5

Photomultiplier and other Detectors

This section helps the reader in developing an improved understanding of photomultipliers. Photomultipliers are extremely sensitive sensors of light and are members of the class of vacuum phototubes. Some of the subjects explained in this segment are photoreceptor cells, resonant-cavity-enhanced photo detectors, microchannel plate detectors and photoelectric sensors.

Photomultiplier

Photomultiplier

Photomultiplier tubes (photomultipliers or PMTs for short), members of the class of vacuum tubes, and more specifically vacuum phototubes, are extremely sensitive detectors of light in the ultraviolet, visible, and near-infrared ranges of the electromagnetic spectrum. These detectors multiply the current produced by incident light by as much as 100 million times (i.e., 160 dB), in multiple dynode stages, enabling (for example) individual photons to be detected when the incident flux of light is very low. Unlike most vacuum tubes, they are not obsolete.

The combination of high gain, low noise, high frequency response or, equivalently, ultra-fast response, and large area of collection has maintained photomultipliers an essential place in nuclear and particle physics, astronomy, medical diagnostics including blood tests, medical imaging, motion picture film scanning (telecine), radar jamming, and high-end image scanners known as drum scanners. Elements of photomultiplier technology, when integrated differently, are the basis of night vision devices.

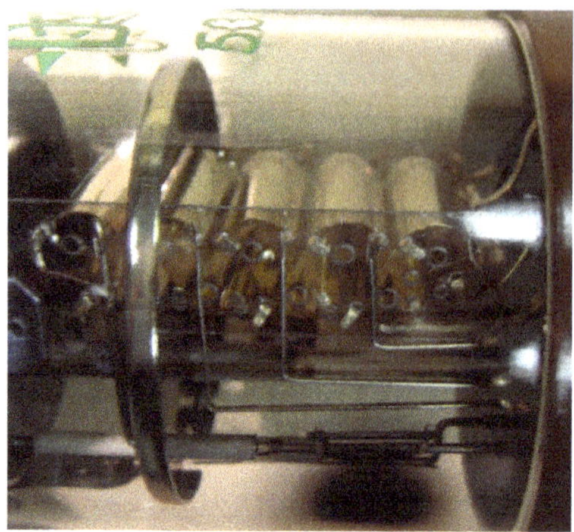

Dynodes inside a photomultiplier tube

Semiconductor devices, particularly avalanche photodiodes, are alternatives to photomultipliers; however, photomultipliers are uniquely well-suited for applications requiring low-noise, high-sensitivity detection of light that is imperfectly collimated.

Structure and Operating Principles

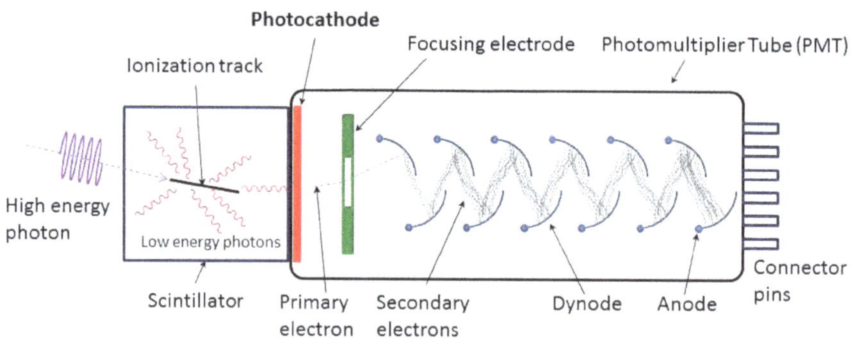

Schematic of a photomultiplier tube coupled to a scintillator.
This arrangement is for detection of gamma rays.

Photomultipliers are typically constructed with an evacuated glass housing, containing a photocathode, several dynodes, and an anode. Incident photons strike the photocathode material, which is usually a thin vapor-deposited conducting layer on the inside of the entry window of the device. Electrons are ejected from the surface as a consequence of the photoelectric effect. These electrons are directed by the focusing electrode toward the electron multiplier, where electrons are multiplied by the process of secondary emission.

The electron multiplier consists of a number of electrodes called *dynodes*. Each dynode is held at a more positive potential, by ≈100 Volts, than the preceding one. A primary electron leaves the photocathode with the energy of the incoming photon, or about 3 eV for "blue" photons, minus the work function of the photocathode. A small group of primary electrons is created by the arrival of a group of initial photons. (In the example Figure, the number of primary electrons in the initial group is proportional to the energy of the incident high energy gamma ray.) The primary electrons move toward the first dynode because they are accelerated by the electric field. They each arrive with ≈100 eV kinetic energy imparted by the potential difference. Upon striking the first dynode, more low energy electrons are emitted, and these electrons are in turn accelerated toward the second dynode. The geometry of the dynode chain is such that a cascade occurs with an exponentially-increasing number of electrons being produced at each stage. For example, if at each stage an average of 5 new electrons are produced for each incoming electron, and if there are 12 dynode stages, then at the last stage one expects for each primary electron about $5^{12} \approx 10^8$ electrons. This last stage is called the anode. This large number of electrons reaching the anode results in a sharp current pulse that is easily detectable, for example on an oscilloscope, signaling the arrival of the photon(s) at the photocathode ≈50 nanoseconds earlier.

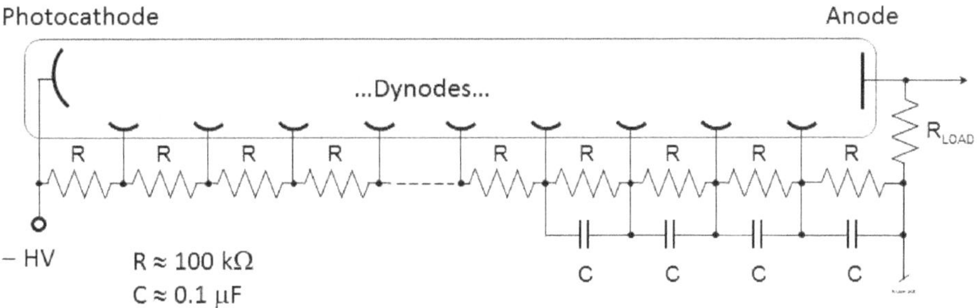

Typical photomultiplier voltage divider circuit using negative high voltage.

The necessary distribution of voltage along the series of dynodes is created by a voltage divider chain, as illustrated in the Figure. In the example, the photocathode is held at a negative high voltage of order 1000V, while the anode is very close to ground potential. The capacitors across the final few dynodes act as local reservoirs of charge to help maintain the voltage on the dynodes while electron avalanches propagate through the tube. Many variations of design are used in practice; the design shown is merely illustrative.

There are two common photomultiplier orientations, the *head-on* or *end-on* (transmission mode) design, as shown above, where light enters the flat, circular top of the tube and passes the photocathode, and the *side-on* design (reflection mode), where light enters at a particular spot on the side of the tube, and impacts on an opaque photocathode. The side-on design is used, for instance, in the type 931, the first mass-produced PMT. Besides the different photocathode materials, performance is also affected by the transmission of the window material that the light passes through, and by the arrangement of the dynodes. A large number of photomultiplier models are available having various combinations of these, and other, design variables. Either of the manuals mentioned will provide the information needed to choose an appropriate design for a particular application.

History

Combining two Scientific Discoveries

The invention of the photomultiplier is predicated upon two prior achievements, the separate discoveries of the photoelectric effect and of secondary emission.

Photoelectric Effect

The first demonstration of the photoelectric effect was carried out in 1887 by Heinrich Hertz using ultraviolet light. Significant for practical applications, Elster and Geitel two years later demonstrated the same effect using *visible* light striking alkali metals (potassium and sodium). The addition of caesium, another alkali metal, has permitted the range of sensitive wavelengths to be extended towards longer wavelengths in the red portion of the visible spectrum.

Historically, the photoelectric effect is associated with Albert Einstein, who relied upon the phenomenon to establish the fundamental principle of quantum mechanics in 1905, an accomplishment for which Einstein received the 1921 Nobel Prize. It is worthwhile to note that Heinrich Hertz, working 18 years earlier, had not recognized that the kinetic energy of the emitted electrons is proportional to the frequency but independent of the optical intensity. This fact implied a discrete nature of light, i.e. the existence of *quanta*, for the first time.

Secondary Emission

The phenomenon of secondary emission (the ability of electrons in a vacuum tube to cause the emission of additional electrons by striking an electrode) was, at first, limited to purely electronic phenomena and devices (which lacked photosensitivity). In 1902, Austin and Starke reported that the metal surfaces impacted by electron beams emitted a larger number of electrons than were incident. The application of the newly discovered secondary emission to the amplification of signals was only proposed after World War I by Westinghouse scientist Joseph Slepian in a 1919 patent.

The Race Towards a Practical Electronic Television Camera

The ingredients for inventing the photomultiplier were coming together during the 1920s as the pace of vacuum tube technology accelerated. The primary goal for many, if not most, workers was the need for a practical television camera technology. Television had been pursued with primitive prototypes for decades prior to the 1934 introduction of the first practical camera (the iconoscope). Early prototype television cameras lacked sensitivity. Photomultiplier technology was pursued to enable television camera tubes, such as the iconoscope and (later) the orthicon, to be sensitive enough to be practical. So the stage was set to combine the dual phenomena of photoemission (i.e., the photoelectric effect) with secondary emission, both of which had already been studied and adequately understood, to create a practical photomultiplier.

First Photomultiplier, Single-stage (Early 1934)

The first documented photomultiplier demonstration dates to the early 1934 accomplishments of an RCA group based in Harrison, NJ. Harley Iams and Bernard Salzberg were the first to integrate a photoelectric-effect cathode and single secondary emission amplification stage in a single vacuum envelope and the first to characterize its performance as a photomultiplier with electron amplification gain. These accomplishments were finalized *prior* to June 1934 as detailed in the manuscript submitted to Proceedings of the Institute of Radio Engineers (Proc. IRE). The device consisted of a semi-cylindrical photocathode, a secondary emitter mounted on the axis, and a collector grid surrounding the secondary emitter. The tube had a gain of about eight and operated at frequencies well above 10 kHz.

Magnetic Photomultipliers (Mid 1934–1937)

Higher gains were sought than those available from the early single-stage photomultipliers. However, it is an empirical fact that the yield of secondary electrons is limited in any given secondary emission process, regardless of acceleration voltage. Thus, any single-stage photomultiplier is limited in gain. At the time the maximum first-stage gain that could be achieved was approximately 10 (very significant developments in the 1960s permitted gains above 25 to be reached using negative electron affinity dynodes). For this reason, multiple-stage photomultipliers, in which the photoelectron yield could be multiplied successively in several stages, were an important goal. The challenge was to cause the photoelectrons to impinge on successively higher-voltage electrodes rather than to travel directly to the highest voltage electrode. Initially this challenge was overcome by using strong magnetic fields to bend the electrons' trajectories. Such a scheme had earlier been conceived by inventor J. Slepian by 1919 (see above). Accordingly, leading international research organizations turned their attention towards improving photomultiplers to achieve higher gain with multiple stages.

In the USSR, RCA-manufactured radio equipment was introduced on a large scale by Joseph Stalin to construct broadcast networks, and the newly formed All-Union Scientific Research Institute for Television was gearing up a research program in vacuum tubes that was advanced for its time and place. Numerous visits were made by RCA scientific personnel to the USSR in the 1930s, prior to the Cold War, to instruct the Soviet customers on the capabilities of RCA equipment and to investigate customer needs. During one of these visits, in September 1934, RCA's Vladimir Zworykin was shown the first multiple-dynode photomultiplier, or *photoelectron multiplier*. This pioneering device was proposed by Leonid A. Kubetsky in 1930 which he subsequently built in 1934. The device achieved gains of 1000x or more when demonstrated in June 1934. The work was submitted for print publication only two years later, in July 1936 as emphasized in a recent 2006 publication of the Russian Academy of Sciences (RAS), which terms it "Kubetsky's Tube." The Soviet device used a magnetic field to confine the secondary electrons and relied on the Ag-O-Cs photocathode which had been demonstrated by General Electric in the 1920s.

By October 1935, Vladimir Zworykin, George Ashmun Morton, and Louis Malter of RCA in Camden, NJ submitted their manuscript describing the first comprehensive experimental and theoretical analysis of a multiple dynode tube — the device later called a *photomultiplier* — to Proc. IRE. The RCA prototype photomultipliers also used an Ag-O-Cs (silver oxide-caesium) photocathode. They exhibited a peak quantum efficiency of 0.4% at 800 nm.

Electrostatic Photomultipliers (1937–present)

Whereas these early photomultipliers used the magnetic field principle, electrostatic photomultipliers (with no magnetic field) were demonstrated by Jan Rajchman of RCA Laboratories in Princeton, NJ in the late 1930s and became the standard for all future commercial photomultipliers. The first mass-produced photomultiplier, the Type 931, was of this design and is still commercially produced today.

Improved Photocathodes

Also in 1936, a much improved photocathode, Cs_3Sb (caesium-antimony), was reported by P. Görlich. The caesium-antimony photocathode had a dramatically improved quantum efficiency of 12% at 400 nm, and was used in the first commercially successful photomultipliers manufactured by RCA (i.e., the 931-type) both as a photocathode and as a secondary-emitting material for the dynodes. Different photocathodes provided differing spectral responses.

Spectral Response of Photocathodes

In the early 1940s, the JEDEC (Joint Electron Devices Engineering Council), an industry committee on standardization, developed a system of designating spectral responses. The philosophy included the idea that the product's user need only be concerned about the response of the device rather than how the device may be fabricated. Various combinations of photocathode and window materials were assigned "S-numbers" (spectral numbers) ranging from S-1 through S-40, which are still in use today. For example, S-11 uses the caesium-antimony photocathode with a lime glass window, S-13 uses the same photocathode with a fused silica window, and S-25 uses a so-called "multialkali" photocathode (Na-K-Sb-Cs, or sodium-potassium-antimony-caesium) that provides extended response in the red portion of the visible light spectrum. No suitable photoemissive surfaces have yet been reported to detect wavelengths longer than approximately 1700 nanometers, which can be approached by a special (InP/InGaAs(Cs)) photocathode.

Role of RCA Corporation

For decades, RCA was responsible for performing the most important work in developing and refining photomultipliers. RCA was also largely responsible for the commercialization of photomultiplers. The company compiled and published an authoritative and widely used *Photomultiplier Handbook*. RCA provided printed copies free upon request. The handbook, which continues to be made available online at no cost by the successors to RCA, is considered to be an essential reference.

Following a corporate break-up in the late 1980s involving the acquisition of RCA by General Electric and disposition of the divisions of RCA to numerous third parties, RCA's photomultiplier business became an independent company.

Lancaster, Pennsylvania Facility

The Lancaster, Pennsylvania facility was opened by the U.S. Navy in 1942 and operated by RCA for the manufacture of radio and microwave tubes. Following World War II, the naval facility was acquired by RCA. *RCA Lancaster*, as it became known, was the base for development and production of commercial television products. In subsequent years other products were added, such as cathode ray tubes, photomultiplier tubes, motion-sensing light control switches, and closed-circuit television systems.

Burle Industries

Burle Industries, as a successor to the RCA Corporation, carried the RCA photomultiplier business forward after 1986, based in the Lancaster, Pennsylvania facility. The 1986 acquisition of RCA by General Electric resulted in the divestiture of the RCA Lancaster New Products Division. Hence, 45 years after being founded by the U.S. Navy, its management team, led by Erich Burlefinger, purchased the division and in 1987 founded Burle Industries.

In 2005, after eighteen years as an independent enterprise, Burle Industries and a key subsidiary were acquired by Photonis, a European holding company Photonis Group. Following the acquisition, Photonis was composed of Photonis Netherlands, Photonis France, Photonis USA, and Burle Industries. Photonis USA operates the former Galileo Corporation Scientific Detector Products Group (Sturbridge, Massachusetts), which had been purchased by Burle Industries in 1999. The group is known for microchannel plate detector (MCP) electron multipliers—an integrated micro-vacuum tube version of photomultipliers. MCPs are used for imaging and scientific applications, including night vision devices.

On 9 March 2009, Photonis announced that it would cease all production of photomultipliers at both the Lancaster, Pennsylvania and the Brive, France plants.

Other Companies

The Japan-based company Hamamatsu Photonics (also known as Hamamatsu) has emerged since the 1950s as a leader in the photomultiplier industry. Hamamatsu, in the tradition of RCA, has published its own handbook, which is available without cost on the company's website. Hamamatsu uses different designations for particular photocathode formulations and introduces modifications to these designations based on Hamamatsu's proprietary research and development.

Photocathode Materials

The photocathodes can be made of a variety of materials, with different properties. Typically the materials have low work function and are therefore prone to thermionic emission, causing noise and dark current, especially the materials sensitive in infrared; cooling the photocathode lowers this thermal noise. The most common photocathode materials are:

- Ag-O-Cs: (Also called S1) Transmission-mode, sensitive from 300–1200 nm. High dark current; used mainly in near-infrared, with the photocathode cooled.
- GaAs:Cs: caesium-activated gallium arsenide. Flat response from 300 to 850 nm, fading towards ultraviolet and to 930 nm.
- InGaAs:Cs: caesium-activated indium gallium arsenide. Higher infrared sensitivity than GaAs:Cs. Between 900–1000 nm much higher signal-to-noise ratio than Ag-O-Cs.
- Sb-Cs: (Also called S11) Caesium-activated antimony. Used for reflective mode photocathodes. Response range from ultraviolet to visible. Widely used.
- Bialkali (Sb-K-Cs, Sb-Rb-Cs): caesium-activated antimony-rubidium or antimony-potassium alloy. Similar to Sb:Cs, with higher sensitivity and lower noise. Can be used for transmission-mode; favorable response to a NaI:Tl scintillator flashes makes them widely used in gamma spectroscopy and radiation detection.
    - High-temperature bialkali (Na-K-Sb): can operate up to 175 °C, used in well logging. Low dark current at room temperature.
- Multialkali (Na-K-Sb-Cs): (Also called S20) Wide spectral response from ultraviolet to near-infrared; special cathode processing can extend range to 930 nm. Used in broadband spectrophotometers.
- Solar-blind (Cs-Te, Cs-I): sensitive to vacuum-UV and ultraviolet. Insensitive to visible light and infrared (Cs-Te has cutoff at 320 nm, Cs-I at 200 nm).

Window Materials

The windows of the photomultipliers act as wavelength filters; this may be irrelevant if the cutoff wavelengths are outside of the application range or outside of the photocathode sensitivity range, but special care has to be taken for uncommon wavelengths.

- Borosilicate glass is commonly used for near-infrared to about 300 nm. Glass with very low content of potassium can be used with bialkali photocathodes to lower the background radiation from the potassium-40 isotope.
- Ultraviolet glass transmits visible and ultraviolet down to 185 nm. Used in spectroscopy.
- Synthetic silica transmits down to 160 nm, absorbs less UV than fused silica. Different thermal expansion than kovar (and than borosilicate glass that's expansion-matched to kovar), a graded seal needed between the window and the rest of the tube. The seal is vulnerable to mechanical shocks.
- Magnesium fluoride transmits ultraviolet down to 115 nm. Hygroscopic, though less than other alkali halides usable for UV windows.

Usage considerations

Photomultiplier tubes typically utilize 1000 to 2000 volts to accelerate electrons within the chain of dynodes. The most negative voltage is connected to the cathode, and the most positive voltage is connected to the anode. Negative high-voltage supplies (with the positive terminal grounded) are often preferred, because this configuration enables the photocurrent to be measured at the low voltage side of the circuit for amplification by subsequent electronic circuits operating at low voltage. However, with the photocathode at high voltage, leakage currents sometimes result in unwanted "dark current" pulses that may affect the operation. Voltages are distributed to the dynodes by a resistive voltage divider, although variations such as active designs (with transistors or diodes) are possible. The divider design, which influences frequency response or rise time, can be selected to suit varying applications. Some instruments that use photomultipliers have provisions to vary the anode voltage to control the gain of the system.

While powered (energized), photomultipliers must be shielded from ambient light to prevent their destruction through overexcitation. In some applications this protection is accomplished mechanically by electrical interlocks or shutters that protect the tube when the photomultiplier compartment is opened. Another option is to add overcurrent protection in the external circuit, so that when the measured anode current exceeds a safe limit, the high voltage is reduced.

If used in a location with strong magnetic fields, which can curve electron paths, steer the electrons away from the dynodes and cause loss of gain, photomultipliers are usually magnetically shielded by a layer of soft iron or mu-metal. This magnetic shield is often maintained at cathode potential. When this is the case, the external shield must also be electrically insulated because of the high voltage on it. Photomultipliers with large distances between the photocathode and the first dynode are especially sensitive to magnetic fields.

Typical Applications

- Photomultipliers were the first electric eye devices, being used to measure interruptions in beams of light.
- Photomultipliers are used in conjunction with scintillators to detect Ionizing radiation by means of hand held and fixed radiation protection instruments, and particle radiation in physics experiments.
- Photomultipliers are used in research laboratories to measure the intensity and spectrum of light-emitting materials such as compound semiconductors and quantum dots.
- Photomultipliers are used as the detector in many spectrophotometers. This allows an instrument design that escapes the thermal noise limit on sensitivity, and which can therefore substantially increase the dynamic range of the instrument.
- Photomultipliers are used in numerous medical equipment designs. For example:
    - blood analysis devices used by clinical medical laboratories, such as flow cytometers, utilize photomultipliers to determine the relative concentration of various components in blood samples, in combination with optical filters and incandescent lamps.
    - an array of photomultipliers is used in a Gamma camera
- Photomultipliers are typically used as the detectors in Flying-spot scanners.

High-sensitivity Applications

After 50 years, during which solid-state electronic components have largely displaced the vacuum tube, the photomultiplier remains a unique and important optoelectronic component. Perhaps its most useful quality is that it acts, electronically, as a nearly perfect current source, owing to the high voltage utilized in extracting the tiny currents associated with weak light signals. There is no Johnson noise associated with photomultiplier signal currents, even though they are greatly amplified, e.g., by 100 thousand times (i.e., 100 dB) or more. The photocurrent still contains shot noise.

Photomultiplier-amplified photocurrents can be electronically amplified by a high-input-impedance electronic amplifier (in the signal path subsequent to the photomultiplier), thus producing appreciable voltages even for nearly infinitesimally small photon fluxes. Photomultipliers offer the best possible opportunity to exceed the Johnson noise for many configurations. The aforementioned refers to measurement of light fluxes that, while small, nonetheless amount to a continuous stream of multiple photons.

For smaller photon fluxes, the photomultiplier can be operated in photon-counting, or Geiger, mode. In Geiger mode the photomultiplier gain is set so high (using high voltage) that a single photo-electron resulting from a single photon incident on the primary surface generates a very large current at the output circuit. However, owing to the avalanche of current, a reset of the photomultiplier is required. In either case, the photomultiplier can detect individual photons. The drawback, however, is that not every photon incident on the primary surface is counted either because of less-than-perfect efficiency of the photomultiplier, or because a second photon can arrive at the photomultiplier during the "dead time" associated with a first photon and never be noticed.

A photomultiplier will produce a small current even without incident photons; this is called the *dark current*. Photon-counting applications generally demand photomultipliers designed to minimise dark current.

Nonetheless, the ability to detect single photons striking the primary photosensitive surface itself reveals the quantization principle that Einstein put forth. Photon counting (as it is called) reveals that light, not only being a wave, consists of discrete particles (i.e., photons).

Photoreceptor Cell

A photoreceptor cell is a specialized type of neuron found in the retina that is capable of phototransduction. The great biological importance of photoreceptors is that they convert light (visible electromagnetic radiation) into signals that can stimulate biological processes. To be more specific, photoreceptor proteins in the cell absorb photons, triggering a change in the cell's membrane potential.

There are currently three known types of photoreceptor cells in mammalian eyes: rods, cones, and photosensitive retinal ganglion cells. The two classic photoreceptor cells are rods and cones, each contributing information used by the visual system to form a representation of the visual world, sight. The rods are narrower than the cones and distributed differently across the retina, but the chemical process in each that supports phototransduction is similar. A third class of mammalian photoreceptor cell was discovered during the 1990s: the photosensitive ganglion cells. These cells do not contribute to sight directly, but are thought to support circadian rhythms and pupillary reflex.

There are major functional differences between the rods and cones. Rods are extremely sensitive, and can be triggered by a single photon. At very low light levels, visual experience is based solely on the rod signal. This explains why colors cannot be seen at low light levels: only one type of photoreceptor cell is active.

Cones require significantly brighter light (i.e., a larger numbers of photons) in order to produce a signal. In humans, there are three different types of cone cell, distinguished by their pattern of response to different wavelengths of light. Color experience is calculated from these three distinct signals, perhaps via an opponent process. The three types of cone cell respond (roughly) to light of short, medium, and long wavelengths. Note that, due to the principle of univariance, the firing of the cell depends upon only the number of photons absorbed. The different responses of the three types of cone cells are determined by the likelihoods that their respective photoreceptor proteins will absorb photons of different wavelengths. So, for example, an L cone cell contains a photoreceptor protein that more readily absorbs long wavelengths of light (i.e., more "red"). Light of a shorter wavelength can also produce the same response, but it must be much brighter to do so.

The human retina contains about 120 million rod cells and 6 million cone cells. The number and ratio of rods to cones varies among species, dependent on whether an animal is primarily diurnal or nocturnal. Certain owls, such as the tawny owl, have a tremendous number of rods in their retinae. In addition, there are about 2.4 million to 3 million ganglion cells in the human visual system, 1 to 2% of them photosensitive. The axons of ganglion cells form the two optic nerves.

The pineal and parapineal glands are photoreceptive in non-mammalian vertebrates, but not in mammals. Birds have photoactive cerebrospinal fluid (CSF)-contacting neurons within the paraventricular organ that respond to light in the absence of input from the eyes or neurotransmitters. Invertebrate photoreceptors in organisms such as insects and molluscs are different in both their morphological organization and their underlying biochemical pathways. Described here are human photoreceptors.

Histology

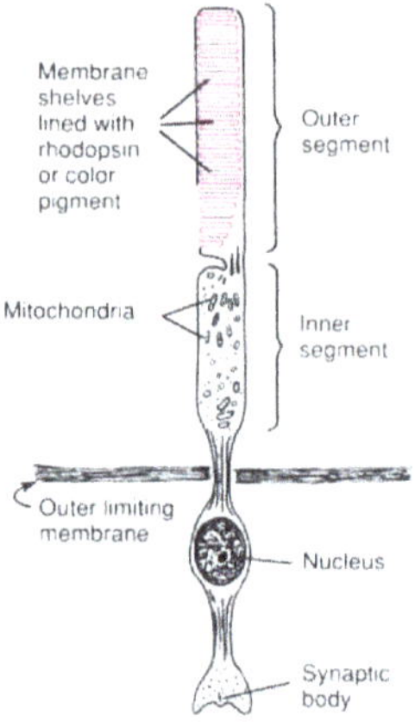

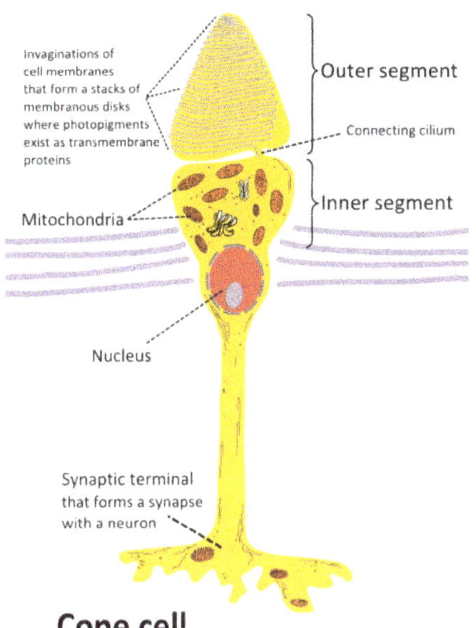

Anatomy of cones and rods varies slightly.

Rod and cone photoreceptors are found on the outermost layer of the retina; they both have the same basic structure. Closest to the visual field (and farthest from the brain) is the axon terminal, which releases a neurotransmitter called glutamate to bipolar cells. Farther back is the cell body, which contains the cell's organelles. Farther back still is the inner segment, a specialized part of the cell full of mitochondria. The chief function of the inner segment is to provide ATP (energy) for the sodium-potassium pump. Finally, closest to the brain (and farthest from the field of view) is the outer segment, the part of the photoreceptor that absorbs light. Outer segments are actually modified cilia that contain disks filled with opsin, the molecule that absorbs photons, as well as voltage-gated sodium channels.

The membranous photoreceptor protein *opsin* contains a pigment molecule called *retinal*. In rod cells, these together are called rhodopsin. In cone cells, there are different types of opsins that combine with retinal to form pigments called photopsins. Three different classes of photopsins in the cones react to different ranges of light frequency, a differentiation that allows the visual system to calculate color. The function of the photoreceptor cell is to convert the light energy of the photon into a form of energy communicable to the nervous system and readily usable to the organism: This conversion is called signal transduction.

The opsin found in the photosensitive ganglion cells of the retina that are involved in various reflexive responses of the brain and body to the presence of (day)light, such as the regulation of circadian rhythms, pupillary reflex and other non-visual responses to light, is called melanopsin. Atypical in vertebrates, melanopsin functionally resembles invertebrate opsins. In structure, it is an opsin, a retinylidene protein variety of G-protein-coupled receptor.

When light activates the melanopsin signaling system, the melanopsin-containing ganglion cells discharge nerve impulses that are conducted through their axons to specific brain targets. These targets include the olivary pretectal nucleus (a center responsible for controlling the pupil of the eye), the LGN, and, through the retinohypothalamic tract (RHT), the suprachiasmatic nucleus of the hypothalamus (the master pacemaker of circadian rhythms). Melanopsin-containing ganglion cells are thought to influence these targets by releasing from their axon terminals the neurotransmitters glutamate and pituitary adenylate cyclase activating polypeptide (PACAP).

Humans

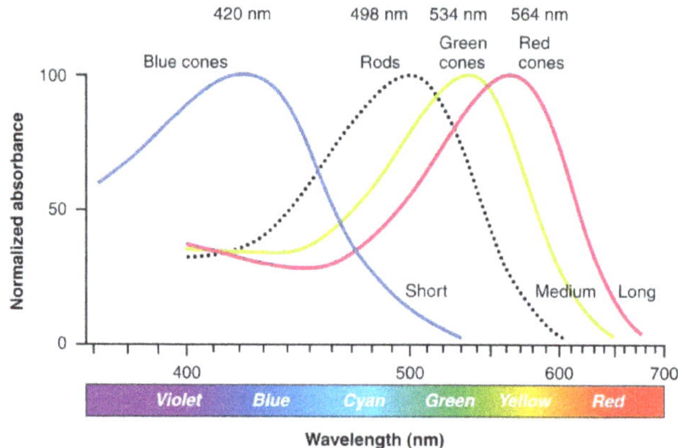

Normalized human photoreceptor absorbances for different wavelengths of light

The human retina has approximately 6 million cones and 120 million rods. Signals from the rods and cones converge on ganglion and bipolar cells for preprocessing before they are sent to the lateral geniculate nucleus. At the "center" of the retina (the point directly behind the lens) lies the fovea (or fovea centralis), which contains only cone cells; and is the region capable of producing the highest visual acuity or highest resolution. Across the rest of the retina, rods and cones are intermingled. No photoreceptors are found at the blind spot, the area where ganglion cell fibers are collected into the optic nerve and leave the eye.

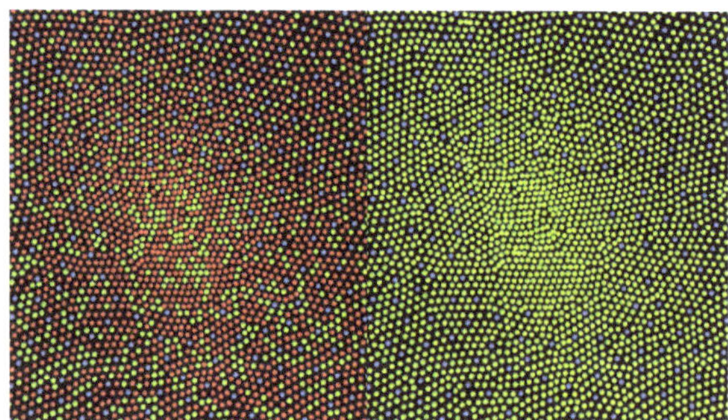
Illustration of the distribution of cone cells in the fovea of an individual with normal color vision (left), and a color blind (protanopic) retina. Note that the center of the fovea holds very few blue-sensitive cones.

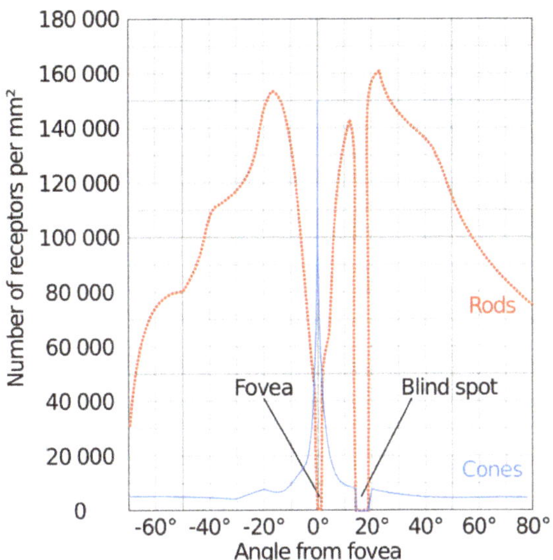

Distribution of rods and cones along a line passing through the fovea and the blind spot of a human eye

The photoreceptor proteins in the three types of cones differ in their sensitivity to photons of different wavelengths (see graph). Since cones respond to both the wavelength and intensity of light, the cone's sensitivity to wavelength is measured in terms of its relative rate of response if the intensity of a stimulus is held fixed, while the wavelength is varied. From this, in turn, is inferred the absorbance. The graph normalizes the degree of absorbance on a hundred-point scale. For example, the S cone's relative response peaks around 420 nm (nanometers, a measure of wavelength). This tells us that an S cone is more likely to absorb a photon at 420 nm than at any other wavelength. If light of a different wavelength to which it is less sensitive, say 480 nm, is increased in brightness appropriately, however, it will produce exactly the same response in the S cone. So, the colors of the curves are misleading. Cones cannot detect color by themselves; rather, color vision requires comparison of the signal across different cone types.

Phototransduction

The process of phototransduction occurs in the *retina*. The *retina* has many layers of various cell types. The best-known *photoreceptor cells* (*rods* and *cones*) form the outermost layer, closest to the sclera, and furthest from the pupil. They are the photoreceptors responsible for sight. The middle layer contains *bipolar cells*, which collect neural signals from the *rods* and the *cones* and then transmit them to the innermost layer of the *retina*, where the *neurons* called *retinal ganglion cells* (RGCs), a small percentage of which are themselves photosensitive, organize the signals and send them to the brain. The bundled *RGC axons* form the *optic nerve*, which leaves the eye through a hole in the *retina* creating the *blind spot*.

Activation of rods and cones is actually hyperpolarization; when they are not being stimulated, they depolarize and release glutamate continuously. In the dark, cells have a relatively high concentration of cyclic guanosine 3'-5' monophosphate (cGMP), which opens ion channels (largely sodium channels, though calcium can enter through these channels as well). The positive charges of the ions that enter the cell down its electrochemical gradient change the cell's membrane potential, cause depolarization, and lead to the release of the neurotransmitter glutamate. Glutamate can depolarize some neurons and hyperpolarize others.

When light hits a photoreceptive pigment within the photoreceptor cell, the pigment changes shape. The pigment, called iodopsin or rhodopsin, consists of large proteins called opsin (situated in the plasma membrane), attached to a covalently bound prosthetic group: an organic molecule called retinal (a derivative of vitamin A). The retinal exists in the 11-cis-retinal form when in the dark, and stimulation by light causes its structure to change to all-trans-retinal. This structural change causes it to activate a regulatory protein called transducin, which leads to the activation of cGMP phosphodiesterase, which breaks cGMP down into 5'-GMP. Reduction in cGMP allows the ion channels to close, preventing the influx of positive ions, hyperpolarizing the cell, and stopping the release of neurotransmitters. The entire process by which light initiates a sensory response is called visual phototransduction.

Dark Current

Unstimulated (in the dark), cyclic-nucleotide gated channels in the outer segment are open because cyclic GMP (cGMP) is bound to them. Hence, positively charged ions (namely sodium ions) enter the photoreceptor, depolarizing it to about −40 mV (resting potential in other nerve cells is usually −65 mV). This depolarizing current is often known as dark current.

Signal Transduction Pathway

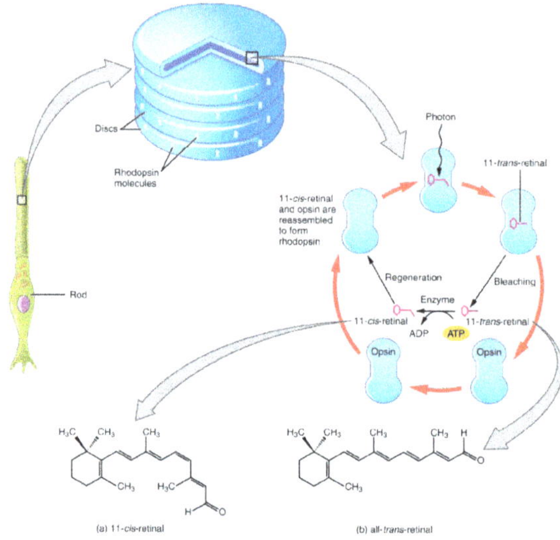

The absorption of light leads to an isomeric change in the retinal molecule.

The signal transduction pathway is the mechanism by which the energy of a photon signals a mechanism in the cell that leads to its electrical polarization. This polarization ultimately leads to either the transmittance or inhibition of a neural signal that will be fed to the brain via the optic nerve. The steps, or signal transduction pathway, in the vertebrate eye's rod and cone photoreceptors are then:

1. The rhodopsin or iodopsin in the disc membrane of the outer segment absorbs a photon, changing the configuration of a retinal Schiff base cofactor inside the protein from the cis-form to the trans-form, causing the retinal to change shape.
2. This results in a series of unstable intermediates, the last of which binds stronger to a G protein in the membrane, called transducin, and activates it. This is the first amplification step – each photoactivated rhodopsin triggers activation of about 100 transducins.
3. Each transducin then activates the enzyme cGMP-specific phosphodiesterase (PDE).
4. PDE then catalyzes the hydrolysis of cGMP to 5' GMP. This is the second amplification step, where a single PDE hydrolyses about 1000 cGMP molecules.
5. The net concentration of intracellular cGMP is reduced (due to its conversion to 5' GMP via PDE), resulting in the closure of cyclic nucleotide-gated Na^+ ion channels located in the photoreceptor outer segment membrane.
6. As a result, sodium ions can no longer enter the cell, and the photoreceptor outer segment membrane becomes hyperpolarized, due to the charge inside the membrane becoming more negative.
7. This change in the cell's membrane potential causes voltage-gated calcium channels to close. This leads to a decrease in the influx of calcium ions into the cell and thus the intracellular calcium ion concentration falls.
8. A decrease in the intracellular calcium concentration means that less glutamate is released via calcium-induced exocytosis to the bipolar cell. (The decreased calcium level slows the release of the neurotransmitter glutamate, which excites the postsynaptic bipolar cells and horizontal cells.)
9. Reduction in the release of glutamate means one population of bipolar cells will be depolarized and a separate population of bipolar cells will be hyperpolarized, depending on the nature of receptors (ionotropic or metabotropic) in the postsynaptic terminal.

Thus, a rod or cone photoreceptor actually releases less neurotransmitter when stimulated by light. Less neurotransmitter could either stimulate (depolarize) or inhibit (hyperpolarize) the bi-polar cell it synapses with, dependent on the nature of the receptor on the bipolar cell. This ability is integral to the center on/off mapping of visual units.

ATP provided by the inner segment powers the sodium-potassium pump. This pump is necessary to reset the initial state of the outer segment by taking the sodium ions that are entering the cell and pumping them back out.

Although photoreceptors are neurons, they do not conduct action potentials with the exception of the photosensitive ganglion cell – which are involved mainly in the regulation of circadian rhythms, melatonin, and pupil dilation.

Advantages

Phototransduction in rods and cones is unique in that the stimulus (in this case, light) actually reduces the cell's response or firing rate, which is unusual for a sensory system where the stimulus usually increases the cell's response or firing rate. However, this system offers several key advantages.

First, the classic (rod or cone) photoreceptor is depolarized in the dark, which means many sodium ions are flowing into the cell. Thus, the random opening or closing of sodium channels will not affect the membrane potential of the cell; only the closing of a large number of channels, through absorption of a photon, will affect it and signal that light is in the visual field. Hence, the system is noiseless.

Second, there is a lot of amplification in two stages of classic phototransduction: one pigment will activate many molecules of transducin, and one PDE will cleave many cGMPs. This amplification means that even the absorption of one photon will affect membrane potential and signal to the brain that light is in the visual field. This is the main feature that differentiates rod photoreceptors from cone photoreceptors. Rods are extremely sensitive and have the capacity of registering a single photon of light, unlike cones. On the other hand, cones are known to have very fast kinetics in terms of rate of amplification of phototransduction, unlike rods.

Difference Between rods and Cones

Comparison of human rod and cone cells, from Eric Kandel et al. in *Principles of Neural Science.*

Rods	Cones
Used for scotopic vision (vision under low light conditions)	Used for photopic vision (vision under high light conditions)
Very light sensitive; sensitive to scattered light	Not very light sensitive; sensitive to only direct light
Loss causes night blindness	Loss causes legal blindness
Low visual acuity	High visual acuity; better spatial resolution
Not present in fovea	Concentrated in fovea
Slow response to light, stimuli added over time	Fast response to light, can perceive more rapid changes in stimuli

Have more pigment than cones, so can detect lower light levels	Have less pigment than rods, require more light to detect images
Stacks of membrane-enclosed disks are unattached to cell membrane directly	Disks are attached to outer membrane
About 120 million rods distributed around the retina	About 6 million cones distributed in each retina
One type of photosensitive pigment	Three types of photosensitive pigment in humans
Confer achromatic vision	Confer color vision

Function

Photoreceptors do not signal color; they only signal the presence of light in the visual field.

A given photoreceptor responds to both the wavelength and intensity of a light source. For example, red light at a certain intensity can produce the same exact response in a photoreceptor as green light of a different intensity. Therefore, the response of a single photoreceptor is ambiguous when it comes to color.

To determine color, the visual system compares responses across a population of photoreceptors (specifically, the three different cones with differing absorption spectra). To determine intensity, the visual system computes how many photoreceptors are responding. This is the mechanism that allows trichromatic color vision in humans and some other animals.

Development

The key events mediating rod versus S cone versus M cone differentiation are induced by several transcription factors, including RORbeta, OTX2, NRL, CRX, NR2E3 and TRbeta2. The S cone fate represents the default photoreceptor program, however differential transcriptional activity can bring about rod or M cone generation. L cones are present in primates, however there is not much known for their developmental program due to use of rodents in research. There are five steps to developing photoreceptors: proliferation of multi-potent retinal progenitor cells (RPCs); restriction of competence of RPCs; cell fate specification; photoreceptor gene expression; and lastly axonal growth, synapse formation and outer segment growth.

Early Notch signaling maintains progenitor cycling. Photoreceptor precursors come about through inhibition of Notch signaling and increased activity of various factors including achaete-scute homologue 1. OTX2 activity commits cells to the photoreceptor fate. CRX further defines the photoreceptor specific panel of genes being expressed. NRL expression leads to the rod fate. NR2E3 further restricts cells to the rod fate by repressing cone genes. RORbeta is needed for both rod and cone development. TRbeta2 mediates the M cone fate. If any of the previously mentioned factors' functions are ablated, the default photoreceptor is a S cone. These events take place at different time periods for different species and include a complex pattern of activities that bring about a spectrum of phenotypes. If these regulatory networks are disrupted, retinitis pigmentosa, macular degeneration or other visual deficits may result.

Signaling

The rod and cone photoreceptors signal their absorption of photons via a decrease in the release of the neurotransmitter glutamate to bipolar cells at its axon terminal. Since the photoreceptor is depolarized in the dark, a high amount of glutamate is being released to bipolar cells in the dark. Absorption of a photon will hyperpolarize the photoreceptor and therefore result in the release of *less* glutamate at the presynaptic terminal to the bipolar cell.

Every rod or cone photoreceptor releases the same neurotransmitter, glutamate. However, the effect of glutamate differs in the bipolar cells, depending upon the type of receptor imbedded in that cell's membrane. When glutamate binds to an ionotropic receptor, the bipolar cell will depolarize (and therefore will hyperpolarize with light as less glutamate is released). On the other hand, binding of glutamate to a metabotropic receptor results in a hyperpolarization, so this bipolar cell will depolarize to light as less glutamate is released.

In essence, this property allows for one population of bipolar cells that gets excited by light and another population that gets inhibited by it, even though all photoreceptors show the same response to light. This complexity becomes both important and necessary for detecting color, contrast, edges, etc.

Further complexity arises from the various interconnections among bipolar cells, horizontal cells, and amacrine cells in the retina. The final result is differing populations of ganglion cells in the retina, a sub-population of which is also intrinsically photosensitive, using the photopigment melanopsin.

Ganglion Cell (Non-rod Non-cone) Photoreceptors

A non-rod non-cone photoreceptor in the eyes of mice, which was shown to mediate circadian rhythms, was discovered in 1991 by Foster *et al.* These neuronal cells, called intrinsically photosensitive retinal ganglion cells (ipRGC), are a small subset (~1–3%) of the retinal ganglion cells located in the inner retina, that is, in front of the rods and cones located in the outer retina. These light sensitive neurons contain a photopigment, melanopsin, which has an absorption peak of the light at a different wavelength (~480 nm) than rods and cones. Beside circadian / behavioral functions, ipRGCs have a role in initiating the pupillary light reflex.

Dennis Dacey with colleagues showed in a species of Old World monkey that giant ganglion cells expressing melanopsin projected to the lateral geniculate nucleus (LGN).

Previously only projections to the midbrain (pre-tectal nucleus) and hypothalamus (suprachiasmatic nucleus) had been shown. However a visual role for the receptor was still unsuspected and unproven.

In 2007, Farhan H. Zaidi and colleagues published pioneering work using rodless coneless humans. *Current Biology* subsequently announced in their 2008 editorial, commentary and despatches to scientists and ophthalmologists, that the non-rod non-cone photoreceptor had been conclusively discovered in humans using landmark experiments on rodless coneless humans by Zaidi and colleagues As had been found in other mammals, the identity of the non-rod non-cone photoreceptor in humans was found to be a ganglion cell in the inner retina. The workers had tracked down patients with rare diseases wiping out classic rod and cone photoreceptor function but preserving ganglion cell function. Despite having no rods or cones the patients continued to exhibit circadian photoentrainment, circadian behavioural patterns, melanopsin suppression, and pupil reactions, with peak spectral sensitivities to environmental and experimental light matching that for the melanopsin photopigment. Their brains could also associate vision with light of this frequency.

In humans the retinal ganglion cell photoreceptor contributes to conscious sight as well as to non-image-forming functions like circadian rhythms, behaviour and pupil reactions. Since these cells respond mostly to blue light, it has been suggested that they have a role in mesopic vision. Zaidi and colleagues' work with rodless coneless human subjects hence also opened the door into image-forming (visual) roles for the ganglion cell photoreceptor. It was discovered that there are parallel pathways for vision – one classic rod and cone-based pathway arising from the outer retina, and the other a rudimentary visual brightness detector pathway arising from the inner retina, which seems to be activated by light before the other. Classic photoreceptors also feed into the novel photoreceptor system, and colour constancy may be an important role as suggested by Foster. The receptor could be instrumental in understanding many diseases including major causes of blindness worldwide like glaucoma, a disease that affects ganglion cells, and the study of the receptor offered potential as a new avenue to explore in trying to find treatments for blindness. It is in these discoveries of the novel photoreceptor in humans and in the receptors role in vision, rather than its non-image-forming functions, where the receptor may have the greatest impact on society as a whole, though the impact of disturbed circadian rhythms is another area of relevance to clinical medicine.

Most work suggests that the peak spectral sensitivity of the receptor is between 460 and 482 nm. Steven Lockley et al. in 2003 showed that 460 nm wavelengths of light suppress melatonin twice as much as longer 555 nm light. However, in more recent work by Farhan Zaidi et al., using rodless coneless humans, it was found that what consciously led to light perception was a very intense 481 nm stimulus; this means that the receptor, in visual terms, enables some rudimentary vision maximally for blue light.

Microchannel Plate Detector

A micro-channel plate (MCP) is a planar component used for detection of particles (electrons, ions and neutrons) and impinging radiation (ultraviolet radiation and X-rays). It is closely related to an electron multiplier, as both intensify single particles or photons by the multiplication of electrons via secondary emission. However, because a microchannel plate detector has many separate channels, it can additionally provide spatial resolution.

Basic Design

A micro-channel plate is a slab made from highly resistive material of typically 2 mm thickness with a regular array of tiny tubes or slots (microchannels) leading from one face to the opposite, densely distributed over the whole surface. The microchannels are typically approximately 10 micrometers in diameter (6 micrometer in high resolution MCPs) and spaced apart by approximately 15 micrometers; they are parallel to each other and often enter the plate at a small angle to the surface (~8° from normal).

Operating Mode

Each microchannel is a continuous-dynode electron multiplier, in which the multiplication takes place under the presence of a strong electric field. A particle or photon that enters one of the channels through a small orifice is guaranteed to hit the wall of the channel due to the channel being at an angle to the plate and thus the angle of impact. The impact starts a cascade of electrons that propagates through the channel, which amplifies the original signal by several orders of magnitude depending on the electric field strength and the geometry of the micro-channel plate. After the cascade, the microchannel takes time to recover (or recharge) before it can detect another signal.

The electrons exit the channels on the opposite side where they are themselves detected by additional means, often simply a single metal anode measuring total current. In some applications each channel is monitored independently to produce an image. Phosphors in combination with photomultiplier tubes have also been used.

Chevron MCP

Most modern MCP detectors consist of two microchannel plates with angled channels rotated 180° from each other producing a chevron (v-like) shape. The angle between the channels reduces ion feedback in the device. In a chevron MCP the electrons that exit the first plate start the cascade in the next plate. The advantage of the chevron MCP over the straight channel MCP is significantly more gain at a given voltage. The two MCPs can either be pressed together or have a small gap between them to spread the charge across multiple channels.

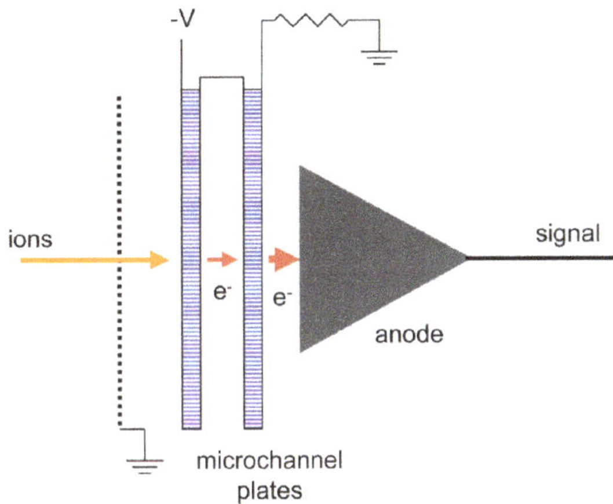

Dual microchannel plate detector schematic

Z stack MCP

This is an assembly of three microchannel plates with channels aligned in a Z shape. Single MCPs can have gain up to 10,000 but this system can provide gain more than 10 million.

The Detector

A microchannel plate within a Finnigan MAT 900 sector mass spectrometer position-and-time-resolved-ion-counting (PATRIC) scanning array detector

An external voltage divider is used to apply 100 volts to the acceleration optics (for electron detection), each MCP, the gap between the MCPs, and the backside of the last MCP and the collector (anode). The last voltage dictates the time of flight of the electrons and in this way the pulse-width. The anode is a 0.4 mm thick plate with an edge of 0.2 mm radius to avoid high field strengths. It is just large enough to cover the active area of the MCP, because the backside of the last MCP and the anode act as a capacitor with 2 mm separation and large capacitance slows down the signal. The positive charge in the MCP influences positive charge in the backside metalization. A hollow torus conducts this around the edge of the anode plate. A torus is the optimum compromise between low capacitance and short path and for similar reasons usually no dielectric (Markor) is placed into this region. After a 90° turn of the torus it is possible to attach a large coaxial waveguide. A taper permits minimizing the radius so that an SMA connector can be used. To save space and make the impedance match less critical, the taper is often reduced to a small 45° cone on the backside of the anode plate.

The typical 500 volts between the backside of the last MCP and the anode cannot be fed into the preamplifier. Therefore, the inner or the outer conductor needs a DC-block, that is, a capacitor. Often it is chosen to only have 10-fold capacitance compared to the MCP-anode capacitance and is implemented as a plate capacitor. Rounded, electro-polished metal plates and the ultra high vacuum allow very high field strengths and high capacitance without a dielectric. The bias for the center conductor is applied via resistors hanging through the waveguide. If the DC block is used in the outer conductor, it is in parallel with the larger capacitor in the power supply. As-suming good screening, the only noise is due to current noise from the linear power regulator. Because the current is low in this application and space for large capacitors is available, and because the DC-block capacitor is fast, it is possible to have very low voltage noise, so that even weak MCP signals can be detected. Sometimes the preamplifier is on a potential (*off ground*) and gets its power through a low-power isolation transformer and outputs its signal optically.

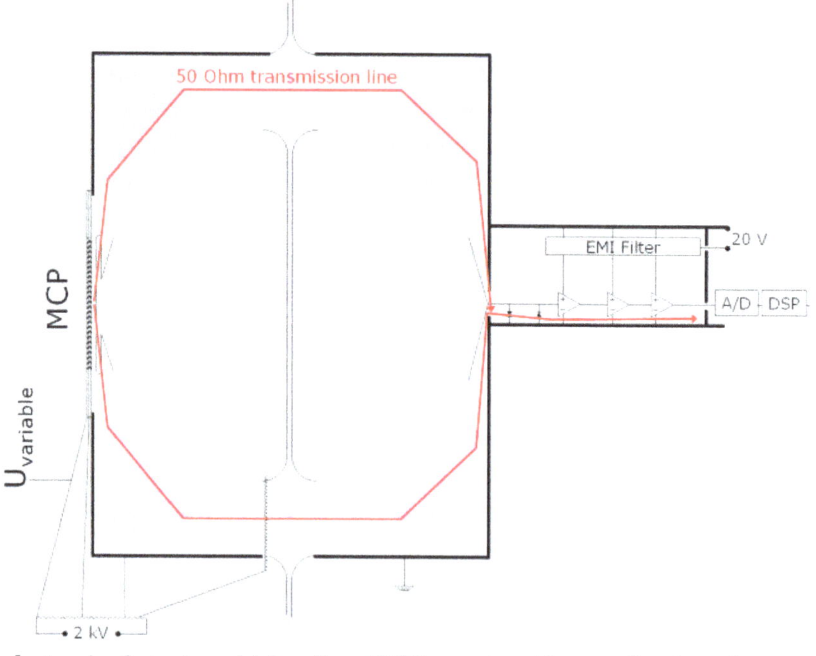

Fast MCP electronics featuring a high voltage UHV capacitor (the grey line from bottom to top)

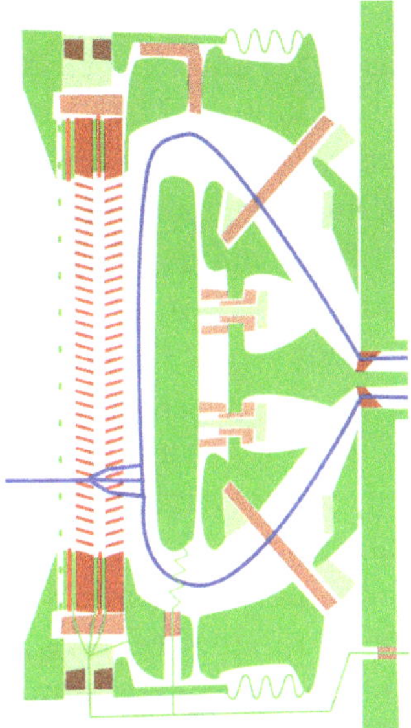

Almost as fast MCP electronics featuring a high voltage UHV capacitor and minimum ceramic

The gain of a MCP is very noisy, especially for single particles. With two thick MCPs (>1 mm) and small channels (< 10 μm), saturation occurs, especially at the ends of the channels after many electron multiplications have taken place. The last stages of the following semiconductor amplifier chain also go into saturation. A pulse of varying length, but stable height and a low jitter leading edge is sent to the time to digital converter. The jitter can be further reduced by means of a constant fraction discriminator. That means that MCP and the preamplifier are used in the linear region (space charge negligible) and the pulse shape is assumed to be due to an impulse response with variable height but fixed shape from a single particle.

Because MCPs have a fixed charge that they can amplify in their life, especially the second MCP has a lifetime problem. It is important to use thin MCPs, low voltage and instead more sensitive and fast semiconductor amplifiers after the anode.

With high count rates or slow detectors (MCPs with phosphor screen or discrete photomultipliers) pulses overlap. In this case a high impedance (slow, but less noisy) amplifier and an ADC is used. Since the output signal from the MCP is generally small, the presence of the thermal noise limits the measurement of the time structure of MCP signal. However, with the fast amplification schemes, is possible to have valuable information on the signal amplitude, even at very low signal values. But yet, not successful on the time structure information of the wideband signals.

Delay Line Detector

In a delay line detector the electrons are accelerated to 500 eV between the back of the last MCP and a grid. Then they fly for 5 mm and are dispersed over an area of 2 mm. A grid follows. Each element has a diameter of 1 mm and consists of an electrostatic lens focusing arriving electrons through a 30 μm hole of a grounded sheet of aluminium. Behind that a cylinder of the same size follows. The electron cloud induces a 300 ps negative pulse when entering the cylinder and a positive when leaving. After that another sheet, a second cylinder follows, and a last sheet follows. Effectively the cylinders are fused into the center-conductor of a stripline. The sheets minimize cross talk between the layers and adjacent lines in the same layer, which would lead to signal dispersion and ringing. These striplines meander across the anode to connect all cylinders, to offer each cylinder 50 Ω impedance, and to generate a position dependent delay. Because the turns in the stripline adversely affect the signal quality their number is limited and for higher resolutions multiple independent striplines are needed. At both ends the meanders are connected to detector electronics. These electronics convert the measured delays into X- (first layer) and Y-coordinates (second layer). Sometimes a hexagonal grid and 3 coordinates are used. This redundancy reduces the dead space-time by reducing the maximum travel distance and thus the maximum delay, allowing for faster measurements. The microchannel plate detector must not operate over around 60 degree Celsius, otherwise it will degrade rapidly, bakeout without voltage has no influence.

Examples of Use

- The mass-market application of microchannel plates is in image intensifier tubes of night vision goggles, which amplify visible and invisible light in order to make dark surroundings visible to the human eye.
- A 1 GHz real-time display CRT for an analog oscilloscope (the Tektronix 7104) used a microchannel plate placed behind the phosphor screen to intensify the image. Without the plate, the image would be excessively dim, because of the electron-optical design.
- MCP detectors are often employed in instrumentation for physical research, and they can be found in devices such as electron and mass spectrometers.

Gaseous Ionization Detectors

Gaseous ionization detectors are radiation detection instruments used in particle physics to detect the presence of ionising particles, and in radiation protection applications to measure ionizing radiation.

They use the ionising effect of radiation upon a gas-filled sensor. If a particle has enough energy to ionize a gas atom or molecule, the resulting electrons and ions cause a current flow which can be measured.

Gaseous ionisation detectors form an important group of instruments used for radiation detection and measurement. This article gives a quick overview of the principal types, and more detailed information can be found in the articles on each instrument. The accompanying plot shows the variation of ion pair generation with varying applied voltage for constant incident radiation. There are three main practical operating regions, one of which each type utilises.

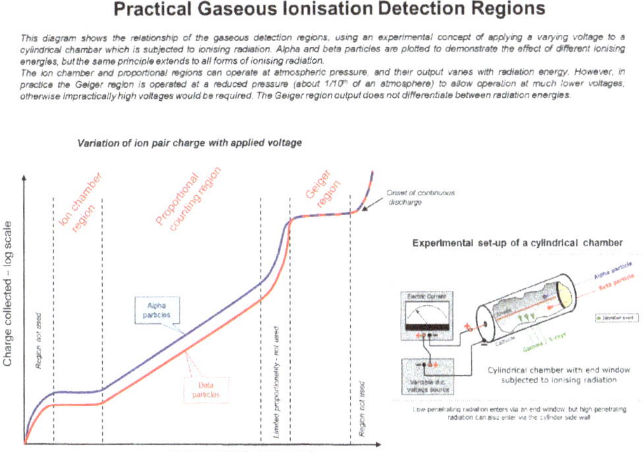

Plot of variation of ion pair current against applied voltage for a wire cylinder gaseous radiation detector.

Types

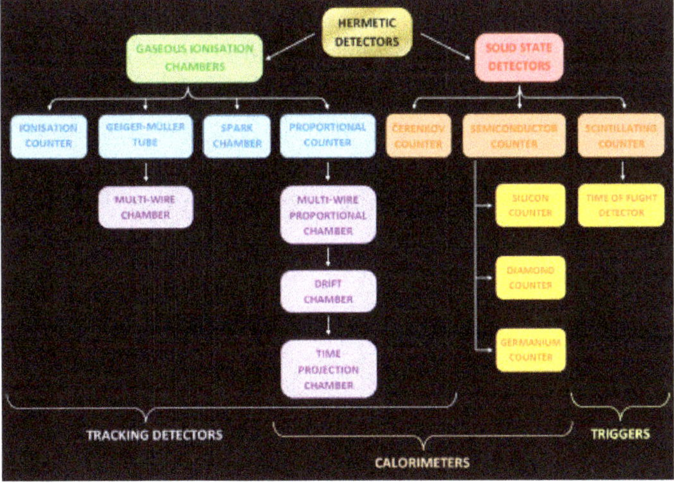

Families of ionising radiation detectors

The three basic types of gaseous ionization detectors are:

- ionization chambers
- proportional counters
- Geiger-Müller tubes

All of these have the same basic design of two electrodes separated by air or a special fill gas, but each uses a different method to measure the total number of ion-pairs that are collected. The strength of the electric field between the electrodes and the type and pressure of the fill gas determines the detector's response to ionizing radiation.

Ionization Chamber

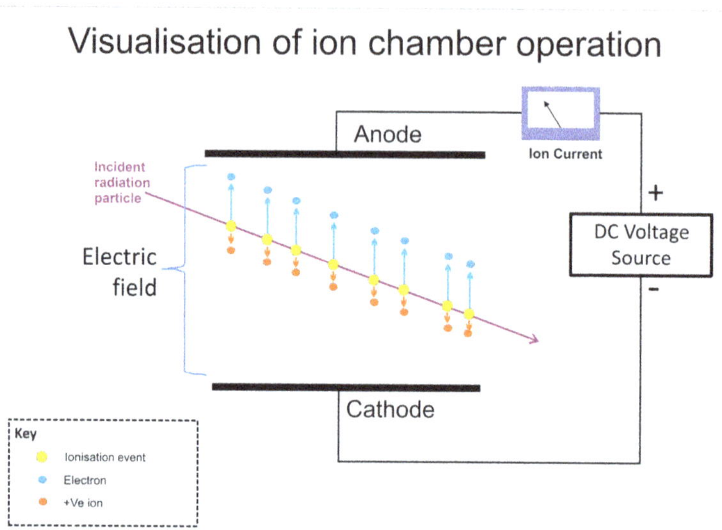

Schematic diagram of ion chamber, showing drift of ions. Electrons typically drift 1000 times faster than positive ions due to their much smaller mass.

Ionization chambers operate at a low electric field strength, selected such that no gas multiplication takes place. The ion current is generated by the creation of "ion pairs", consisting of an ion and an electron. The ions drift to the cathode whilst free electrons drift to the anode under the influence of the electric field. This current is independent of the applied voltage if the device is being operated in the "ion chamber region". Ion chambers are preferred for high radiation dose rates because they have no "dead time"; a phenomenon which affects the accuracy of the Geiger Muller tube at high dose rates.

The advantages are:

- Good uniform response to gamma radiation and give an accurate overall dose reading
- Will measure very high radiation rates
- Sustained high radiation levels do not degrade fill gas

The disadvantages are:

- Very low electronic output requiring sophisticated electrometer circuit
- Operation and accuracy easily affected by moisture

Proportional Counter

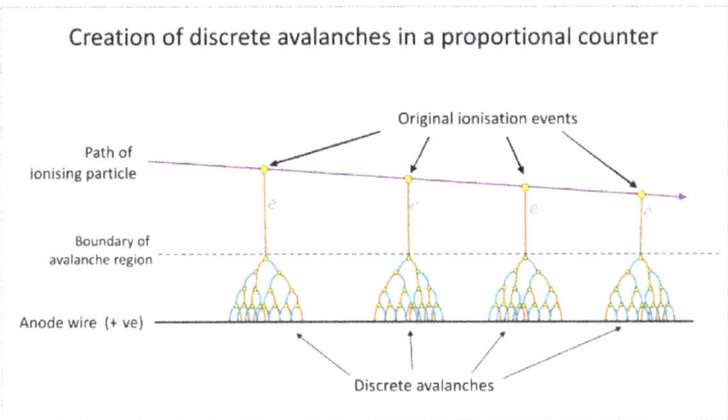

The generation of discrete Townsend avalanches in a proportional counter.

Proportional counters operate at a slightly higher voltage, selected such that discrete avalanches are generated. Each ion pair produces a single avalanche so that an output current pulse is generated which is proportional to the energy deposited by the radiation. This is in the "proportional counting" region. The term "gas proportional detector" (GPD) is generally used in radiometric practice, and the property of being able to detect particle energy is particularly useful when using large area flat arrays for alpha and beta particle detection and discrimination, such as in installed personnel monitoring equipment. The Wire chamber is a multi-electrode form of proportional counter used as a research tool.

The advantages are:

- Can measure energy of radiation and provide spectrographic information
- Can discriminate between alpha and beta particles
- Large area detectors can be constructed

The disadvantages are:

- Anode wires delicate and can lose efficiency in gas flow detectors due to deposition
- Efficiency and operation affected by ingress of oxygen into fill gas
- Measurement windows easily damaged in large area detectors

Geiger-Müller Tube

Geiger-Müller tubes are the primary components of Geiger counters. They operate at an even higher voltage, selected such that each ion pair creates an avalanche, but by the emission of UV photons, multiple avalanches are created which spread along the anode wire, and the adjacent gas volume ionizes from as little as a single ion pair event. This is the "Geiger region" of operation. The current pulses produced by the ionising events are passed to processing electronics which can derive a visual display of count rate or radiation dose, and usually in the case of hand-held instruments, an audio device producing clicks.

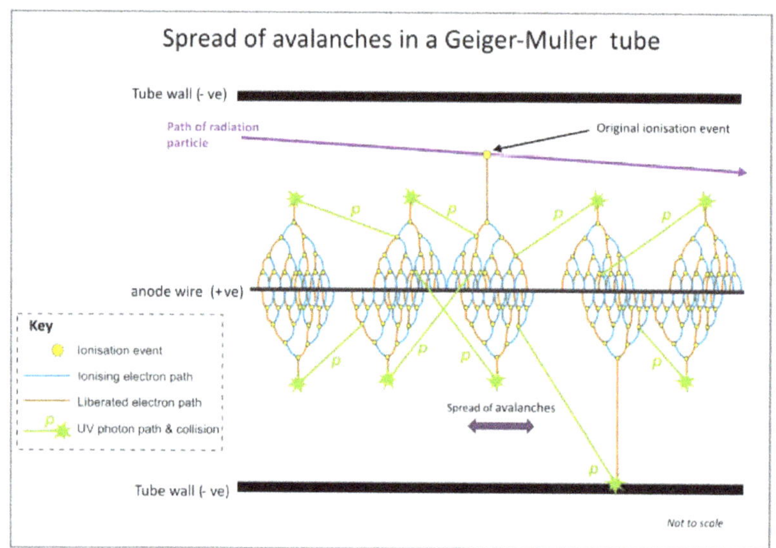

Visualisation of the spread of Townsend avalanches by means of UV photons

The advantages are:

- Cheap, robust detector with a large variety of sizes and applications
- Large output signal from tube requiring minimal electronic processing for simple counting
- Can measure overall gamma dose when using energy compensated tube

The disadvantages are:

- Cannot measure energy of radiation - no spectrographic information
- Will not measure high radiation rates due to dead time
- Sustained high radiation levels will degrade fill gas

Guidance on Detector Type Usage

The UK Health and Safety Executive has issued a guidance note on the correct portable instrument for the application concerned . This covers all radiation instrument technologies and is useful in selecting the correct gaseous ionisation detector technology for a measurement application.

Everyday Use

Ionization-type smoke detectors are gaseous ionization detectors in widespread use. A small source of radioactive americium is placed so that it maintains a current between two plates that effectively form an ionisation chamber. If smoke gets between the plates where ionization is taking place, the ionized gas can be neutralized leading to a reduced current. The decrease in current triggers a fire alarm.

Cryogenic Particle Detectors

Cryogenic particle detectors operate at very low temperature, typically only a few degrees above absolute zero. These sensors interact with an energetic elementary particle (such as a photon) and deliver a signal which can be related to the type of particle and the nature of the interaction. While many types of particle detectors might be operated with improved performance at cryogenic temperatures, this term generally refers to types which take advantage of special effects or properties occurring only at low temperature.

Introduction

The most commonly cited reason for operating any sensor at low temperature is the reduction in thermal noise, which is proportional to the square root of the absolute temperature. However, at very low temperature, certain material properties become very sensitive to energy deposited by particles in their passage through the sensor, and the gain from these changes may be even more than that from reduction in thermal noise. Two such commonly used properties are heat capacity and electrical resistivity, particularly superconductivity; other designs are based on superconducting tunnel junctions, quasiparticle trapping, rotons in superfluids, magnetic bolometers, and other principles.

Originally, astronomy pushed the development of cryogenic detectors for optical and infrared radiation. Later, particle physics and cosmology motivated cryogenic detector development for sensing known and predicted particles such as neutrinos, axions, and weakly interacting massive particles (WIMPs).

Types of Cryogenic Particle Detectors

Calorimetric Particle Detection

A calorimeter is a device which measures the amount of heat deposited in a sample of material. A calorimeter differs from a bolometer in that a calorimeter measures energy while a bolometer measures power.

Below the Debye temperature of a crystalline dielectric material (such as silicon), the heat capacity decreases inversely as the cube of the absolute temperature. It becomes very small so that the sample's increase in temperature for a given heat input may be relatively large. This makes it practical to make a calorimeter that has a very large temperature excursion for a small amount of heat input, such as that deposited by a passing particle. The temperature rise can be measured with a standard type of thermistor, as in a classical calorimeter. In general, small sample size and very sensitive thermistors are required to make a sensitive particle detector by this method.

In principle, several types of resistance thermometers can be used. The limit of sensitivity to energy deposition is determined by the magnitude of resistance fluctuations, which are in turn determined by thermal fluctuations. Since all resistors exhibit voltage fluctuations that are proportional to their temperature, an effect known as Johnson noisc, a reduction of temperature is often the only way to achieve the required sensitivity.

Superconducting Transition Edge Sensors

A very sensitive calorimetric sensor known as a transition edge sensor (TES) takes advantage of superconductivity. Most pure superconductors have a very sharp transition from normal resistivity to superconductivity at some low temperature. By operating on the superconducting phase transition, a very small change in temperature resulting from interaction with a particle results in a significant change in resistance.

Superconducting Tunnel Junctions

The superconducting tunnel junction (STJ) consists of two pieces of superconducting material separated by a very thin (~nanometer) insulating layer. It is also known as a superconductor-insulator-superconductor tunnel junction (SIS), and it is a type a Josephson junction. Cooper pairs can tunnel across the insulating barrier, a phenomenon known as the Josephson effect. Quasiparticles can also tunnel across the barrier, although the quasiparticle current is suppressed for voltages less than twice the superconducting energy gap. A photon absorbed on one side of a STJ breaks Cooper pairs and creates quasiparticles. In the presence of an applied voltage across the junction, the quasiparticles tunnel across the junction, and the resulting tunneling current is proportional to the photon energy. The STJ can also be used as a heterodyne detector by exploiting the change in the nonlinear current-voltage characteristic that results from photon-assisted tunneling. STJs are the most sensitive heterodyne detectors available for the 100 GHz – 1 THz frequency range and are employed for astronomical observation at these frequencies.

Kinetic Inductance Detectors

The kinetic inductance detector (KID) is based on measuring the change in kinetic inductance caused by the absorption of photons in a thin strip of superconducting material. The change in inductance is typically measured via the change in the resonant frequency of a microwave resonator, and hence these detectors are also known as microwave kinetic inductance detectors (MKIDs).

Superconducting Granules

The superconducting transition alone can be used to directly measure the heating caused by a passing particle. A type I superconducting grain in a magnetic field exhibits perfect diamagnetism and excludes the field completely excluded from its interior. If it is held slightly below the transition temperature, the superconductivity vanishes on heating by particle radiation, and the field suddenly penetrates the interior. This field change can be detected by a surrounding coil. The change is reversible when the grain cools again. In practice the grains must be very small and carefully made, and carefully coupled to the coil.

Magnetic Calorimeters

Paramagnetic rare earth ions are being used as particle sensors by sensing the spin flips of the paramagnetic atoms induced by heat absorbed in a low heat capacity material. The ions are used as a magnetic thermometer.

Other Methods

Phonon Particle Detection

Calorimeters assume the sample is in thermal equilibrium or nearly so. In crystalline materials at very low temperature this is not necessarily the case. A good deal more information can be found by measuring the elementary excitations of the crystal lattice, or phonons, caused by the interacting particle. This can be done by several methods including superconducting transition edge sensors.

Superconducting Nanowire Single-photon Detectors

The superconducting nanowire single-photon detector (SNSPD) is based on a superconducting wire cooled well below the superconducting transition temperature and biased with a dc current that is close to but less than the superconducting critical current. The SNSPD is typically made from ≈ 5 nm thick niobium nitride films which are patterned as narrow nanowires (with a typical width of 100 nm). Absorption of a photon breaks Cooper pairs and reduces the critical current below the bias current. A small non-superconducting section across the width of the nanowire is formed. This resistive non-superconducting section then leads to a detectable voltage pulse of a duration of about 1 nanosecond. The main advantages of this type of photon detector are its high speed (a maximal count rate of 2 GHz makes them the fastest available) and its low dark count rate. The main disadvantage is the lack of intrinsic energy resolution.

Roton Detectors

In superfluid ^{4}He the elementary collective excitations are phonons and rotons. A particle striking an electron or nucleus in this superfluid can produce rotons, which may be detected bolometrically or by the evaporation of helium atoms when they reach a free surface. ^{4}He is intrinsically very pure so the rotons travel ballistically and are stable, so that large volumes of fluid can be used.

Quasiparticles in Superfluid ^{3}He

In the B phase, below 0.001 K, superfluid ^{3}He acts similarly to a superconductor. Pairs of atoms are bound as quasiparticles similar to Cooper pairs with a very small energy gap of the order of 100 nanoelectronvolts. This allows building a detector analogous to a superconducting tunnel detector. The advantage is that many (~10^9) pairs could be produced by a single interaction, but the difficulties are that it is difficult to measure the excess of normal ^{3}He atoms produced and to prepare and maintain much superfluid at such low temperature.

Quantum Well Infrared Photodetector

A Quantum Well Infrared Photodetector (QWIP) is an infrared photodetector, which uses electronic intersubband transitions in quantum wells to absorb photons. The basic elements of a QWIP are quantum wells, which are separated by barriers. The quantum wells are designed to have one confined state inside the well and a first excited state which aligns with the top of the barrier. The wells are n-doped such that the ground state is filled with electrons. The barriers are wide enough to prevent quantum tunneling between the quantum wells. Typical QWIPs consists of 20 to 50 quantum wells. When a bias voltage is applied to the QWIP, the entire conduction band is tilted. Without light the electrons in the quantum wells just sit in the ground state. When the QWIP is illuminated with light of the same or higher energy as the intersubband transition energy, an electron is excited.

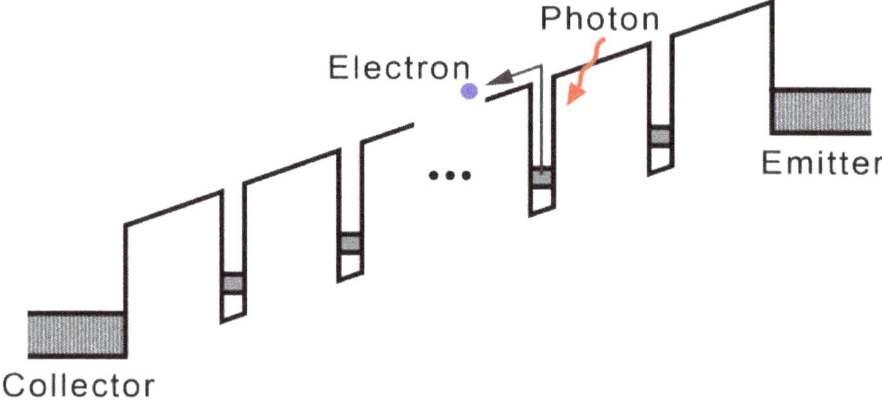

Conduction band profile of a photoconductive QWIP. The conduction band profile is tilted as a bias voltage is applied.

Once the electron is in an excited state, it can escape into the continuum and be measured as photocurrent. To externally measure a photocurrent the electrons need to be extracted by applying an electric field to the quantum wells. The efficiency of this absorption and extraction process depends on several parameters.

Photocurrent

Assuming that the detector is illuminated with a photon flux ϕ (number of photons per unit time), the photocurrent I_{ph} is

$$I_{ph} = e\phi\eta g_{ph}$$

where e is the elementary charge, η is the absorption efficiency and g_{ph} is the photoconductive gain. and g_{ph} are the probabilities for a photon to add an electron to the photocurrent, also called quantum efficiency. η is the probability of a photon exciting an electron, and g_{ph} depends on the electronic transport properties.

Photoconductive Gain

The photoconductive gain g_{ph} is the probability that an excited electron contributes to the photocurrent—or more generally, the number of electrons in the external circuit, divided by the number of quantum well electrons that absorb a photon. Although it might be counterintuitive at first, it is possible for g_{ph} to be larger than one. Whenever an electron is excited and extracted as photocurrent, an extra electron is injected from the opposite (emitter) contact to balance the loss of electrons from the quantum well. In general the capture probability $p_c \leq 1$, so an injected electron might sometimes pass over the quantum well and into the opposite contact. In that case, yet another electron is injected from the emitter contact to balance the charge, and again heads towards the well where it might or might not get captured, and so on, until eventually an electron is captured in the well. In this way, g_{ph} can become larger than one.

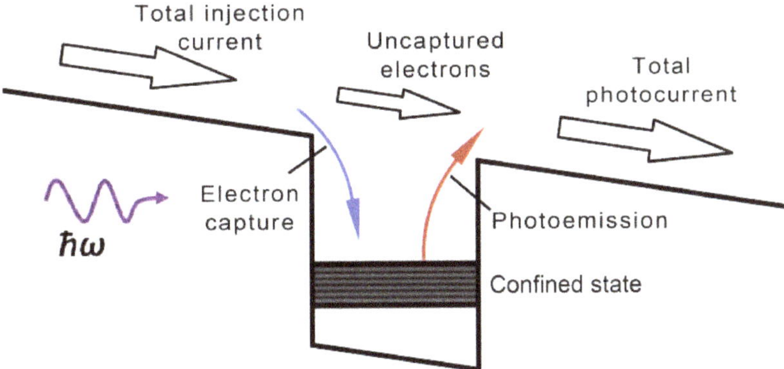

Photoconductive gain in a quantum well infrared photodetector. To balance the loss of electrons from the quantum well, electrons are injected from the top emitter contact. Since the capture probability is smaller than one, extra electrons need to be injected and the total photocurrent can become larger than the photoemission current.

The exact value of g_{ph} is determined by the ratio of capture probability p_c and escape probability p_e.

$$g_{ph} = \frac{p_e}{N p_c}$$

where N is the number of quantum wells. The number of quantum wells appears only in the denominator, as it increases the capture probability p_c, but not the escape probability p_e.

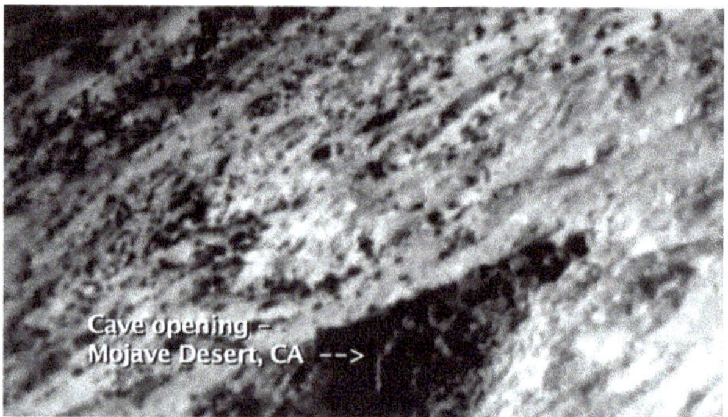

This video shows the evolution of taking the quantum-well infrared photodetector (QWIP) from inception, to testing on the ground and from a plane, and ultimately to a NASA science mission.

Photoelectric Sensor

A photoelectric sensor, or photo eye, is an equipment used to discover the distance, absence, or presence of an object by using a light transmitter, often infrared, and a photoelectric receiver. They are largely used in industrial manufacturing. There are three different useful types: opposed (through beam), retro-reflective, and proximity-sensing (diffused).

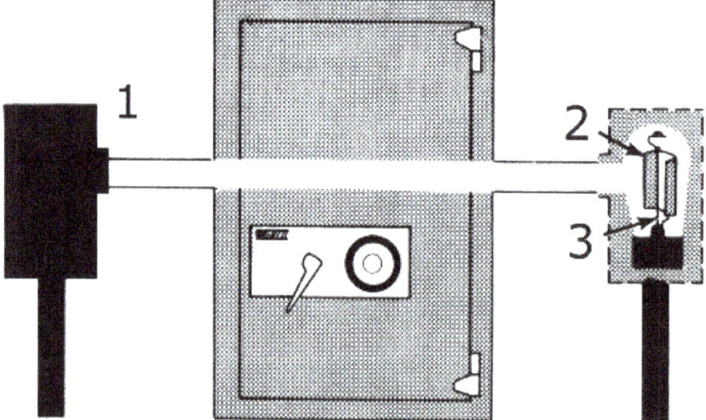

Conceptual through-beam system to detect unauthorized access to a secure door. If the beam is damaged, the detector triggers an alarm .

Types

A self-contained photoelectric sensor contains the optics, along with the electronics. It requires only a power source. The sensor performs its own modulation, demodulation, amplification, and output switching. Some self-contained sensors provide such options as built-in control timers or counters. Because of technological progress, self-contained photoelectric sensors have become increasingly smaller.

Remote photoelectric sensors used for remote sensing contain only the optical components of a sensor. The circuitry for power input, amplification, and output switching are located elsewhere, typically in a control panel. This allows the sensor, itself, to be very small. Also, the controls for the sensor are more accessible, since they may be bigger.

When space is restricted or the environment too hostile even for remote sensors, fiber optics may be used. Fiber optics are passive mechanical sensing components. They may be used with either remote or self-contained sensors. They have no electrical circuitry and no moving parts, and can safely pipe light into and out of hostile environments.

Sensing Modes

A through beam arrangement consists of a receiver located within the line-of-sight of the transmitter. In this mode, an object is detected when the light beam is blocked from getting to the receiver from the transmitter.

A retroreflective arrangement places the transmitter and receiver at the same location and uses a reflector to bounce the light beam back from the transmitter to the receiver. An object is sensed when the beam is interrupted and fails to reach the receiver.

A proximity-sensing (diffused) arrangement is one in which the transmitted radiation must reflect off the object in order to reach the receiver. In this mode, an object is detected when the receiver sees the transmitted source rather than when it fails to see it. As in retro-reflective sensors, diffuse sensor emitters and receivers are located in the same housing. But the target acts as the reflector, so that detection of light is reflected off the disturbance object. The emitter sends out a beam of light (most often a pulsed infrared, visible red, or laser) that diffuses in all directions, filling a detection area. The target then enters the area and deflects part of the beam back to the receiver. Detection occurs and output is turned on or off when sufficient light falls on the receiver.

Some photo eyes have two different operational types, light operate and dark operate. Light operate photo eyes become operational when the receiver "receives" the transmitter signal. Dark operate photo eyes become operational when the receiver "does not receive" the transmitter signal.

The detecting range of a photoelectric sensor is its "field of view", or the maximum distance from which the sensor can retrieve information, minus the minimum distance. A minimum detectable object is the smallest object the sensor can detect. More accurate sensors can often have minimum detectable objects of minuscule size.

Certain types of smoke detector use a photoelectric sensor to warn of smoldering fires.

Difference Between Modes

Name	Advantages	Disadvantages
Through-Beam	• Most accurate • Longest sensing range • Very reliable	• Must install at two points on system: emitter and receiver • Costly - must purchase both emitter and receiver
Reflective	• Only slightly less accurate than through-beam • Sensing range better than diffuse • Very reliable	• Must install at two points on system: sensor and reflector • Slightly more costly than diffuse • Sensing range less than through-beam
Diffuse	• Only install at one point • Cost less than through-beam or reflective	• Less accurate than through- beam or reflective • More setup time involved

Superconducting Nanowire Single-photon Detector

The superconducting nanowire single-photon detector (SNSPD) is a type of near-infrared and optical single-photon detector based on a current-biased superconducting nanowire. It was first developed by scientists at Moscow State Pedagogical University and at the University of Rochester in 2001.

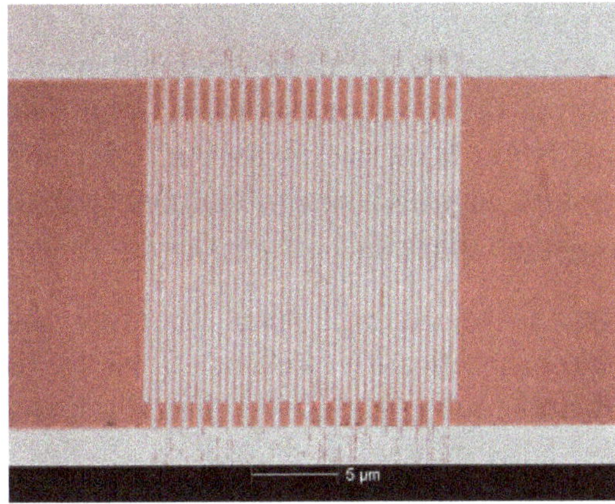

False-color scanning electron micrograph of a superconducting nanowire single-photon detector (SNSPD). Image credit: NIST.

As of 2013, a superconducting nanowire single-photon detector is the fastest single-photon detector (SPD) for photon counting.

Principle of Operation

The SNSPD consists of a thin (≈ 5 nm) and narrow (≈ 100 nm) superconducting nanowire. The length is typically hundreds of micrometers, and the nanowire is patterned in a compact meander geometry to create a square or circular pixel with high detection efficiency. The nanowire is cooled well below its superconducting critical temperature and biased with a DC current that is close to but less than the superconducting critical current of the nanowire. A photon incident on the nanowire breaks Cooper pairs and reduces the local critical current below that of the bias current. This results in the formation of a localized non-superconducting region, or hotspot, with finite electrical resistance. This resistance is typically larger than the 50 ohm input impedance of the readout amplifier, and hence most of the bias current is shunted to the amplifier. This produces a measurable voltage pulse that is approximately equal to the bias current multiplied by 50 ohms. With most of the bias current flowing through the amplifier, the non-superconducting region cools and returns to the superconducting state. The time for the current to return to the nanowire is typically set by the inductive time constant of the nanowire, equal to the kinetic inductance of the nanowire divided by the impedance of the readout circuit. Proper self-resetting of the device requires that this inductive time constant be slower than the intrinsic cooling time of the nanowire hotspot.

While the SNSPD does not offer the intrinsic energy or photon-number resolution of the superconducting transition edge sensor, the SNSPD is significantly faster than conventional transition edge sensors and operates at higher temperatures. Most SNSPDs are made of niobium nitride (NbN), which offers a relatively high superconducting critical temperature (≈ 10 K) and a very fast cooling time (<100 picoseconds). NbN devices have demonstrated device detection efficiencies as high as 67% at 1064 nm wavelength with count rates in the hundreds of MHz. NbN devices have also demonstrated jitter

– the uncertainty in the photon arrival time – of less than 50 picoseconds, as well as very low rates of dark counts, i.e. the occurrence of voltage pulses in the absence of a detected photon. In addition, the deadtime (time interval following a detection event during which the detector is not sensitive) is on the order of a few nanoseconds, this short deadtime translates into very high saturation count rates and enables antibunching measurements with a single detector.

For the detection of longer wavelength photons, however, the detection efficiency of standard SNSPDs decreases significantly. Recent efforts to improve the detection efficiency at near-infrared and mid-infrared wavelengths include studies of narrower (20 nm and 30 nm wide) NbN nanowires as well as studies of materials with lower superconducting critical temperatures than NbN (tungsten silicide, niobium silicide, and tantalum nitride).

Applications

Many of the initial application demonstrations of SNSPDs have been in the area of quantum information, such as quantum key distribution and quantum computing. Other applications include imaging of infrared photoemission for defect analysis in CMOS circuitry, LIDAR, on-chip quantum optics, single plasmon detection, quantum plasmonics, single electron detection, single α and β particles detection, oxygen singlet luminescence detection and ultra-long distance classical communication. A number of companies are commercializing complete single photon detection systems based on superconducting nanowires, including Scontel, Photon Spot, Single Quantum, and Quantum Opus.

Resonant-Cavity-Enhanced Photo Detector

Resonant-cavity-enhanced photo detectors (or, RCE photo detectors) enable improved performance over their predecessors by placing the active device structure inside a Fabry–Pérot resonant cavity. Though the active device structure of the RCE detectors remains close to other conventional photodetectors, the effect of the optical cavity, which allows wavelength selectivity and an enhancement of the optical field due to resonance, allows the photo detectors to be made thinner and therefore faster, while simultaneously increasing the quantum efficiency at the resonant wavelengths.

Advantages

The quantum efficiency of conventional detectors is dominated by the optical absorption (electromagnetic radiation) of the semiconductor material. For semiconductors with low absorption coefficients, a thicker absorption regions is required to achieve higher quantum efficiency, but at the cost of the Bandwidth (signal processing) of photodetectors.

A RCE detector improves the bandwidth significantly. The constructive interference of a Fabry–Perot cavity enhances the optical field inside the photodetector at the resonance wavelengths to achieve a quantum efficiency of close to unity. Moreover the optical cavity makes the RCE detectors wavelength selective. This makes RCE photodetectors attractive for low crosstalk wavelength demultiplexing. Improved quantum efficiency gives less power consumption. Higher bandwidth gives faster operation.

The RCE photodetectors have both wavelength selectivity and high speed response making them ideal for wavelength division multiplexing applications. Optical modulators situated in an optical cavity require fewer quantum wells to absorb the same fraction of the incident light, and can therefore operate at lower voltages. In the case of emitters, the cavity modifies the spontaneous emission of light-emitting diodes (LED) improving their spectral purity and directivity.

Thus optical communication systems can perform much faster, with more bandwidth and can become more reliable. Camera sensors could give more resolutions, better contrast ratios and less distortion. For these reasons, RCE devices can be expected to play a growing role in optoelectronics over the coming years.

Theory of RCE Photo Detectors

The RCE photo detectors can provide:

1. Higher quantum efficiency,
2. Higher detection speed,
3. Wavelength selective detection,

than compare to a conventional photodiode.

Quantum Efficiency of RCE Photo Detectors

The RCE photodetectors are expected to have higher quantum efficiency η than compare to conventional photodiodes. The formulation of η for RCE devices gives insight to the design criteria.

A generalized RCE photodetector schematic as given in Figure 1 can give the required theoretical model of photodetection. A thin absorption region of thickness d is sandwiched between two relatively less absorbing region, substrate, of thickness L_1 and L_2. The optical cavity is formed by a period of $\lambda/4$ distributed Bragg reflector (DBR), made of non-absorbing larger bandgap materials, at the both end of the substrate. The front mirror has a transmittance of t_1 and generally has lower reflectivity than compare to the mirror at back ($R_1 < R_2$). Transmittance t_1 allows light to enter into the cavity, and reflectivity R_1 ($=r_1^2$) and R_2 ($=r_2^2$) provides the optical confinement in the cavity.

The active region and the substrate region have absorption coefficient α and α_{ex} respectively. The field reflection coefficients of the front and the back mirrors are $r_1e^{-j\phi_1}$ and $r_2e^{-j\phi_2}$ respectively, where ϕ_1 and ϕ_2 are the phase shifts due to the light penetration into the mirrors.

The optical microcavity allows building up an optical field inside the optical cavity. In compare to conventional detector, where light is absorbed in a single pass through the absorption region, for RCE detectors trapped light is absorbed each time it traverses through the absorption region.

The Quantum efficiency η for a RCE detector is given by:

$$\eta = (1-R_1)(1-e^{-\alpha d})[\frac{(e^{-\alpha_{ex}L_1} + r_2^2 e^{-\alpha_{ex}L_2-\alpha_c L})}{1-2r_1r_2e^{-\alpha_c L}\cos(2\beta L+\phi_1+\phi_2)+(r_1r_2)^2 e^{-\alpha_c L}}]$$

Here $\alpha_c = (\alpha_{ex}L_1 + \alpha_{ex}L_2 + \alpha d)/L$. In practical detector design $\alpha_{ex} << \alpha$, so α_{ex} can be neglected and η can be given as:

$$\eta = (1-R_1)(1-e^{-\alpha d})[\frac{(1+R_2e^{-\alpha d})}{1-2\sqrt{R_1R_2}e^{-\alpha_c d}\cos(2\beta L+\phi_1+\phi_2)+(R_1R_2)e^{-\alpha_c d}}]$$

The term inside the [] represents the cavity enhancement effect. This is a periodic function of $2\beta L+\phi_1+\phi_2$, which has minima at $2\beta L+\phi_1+\phi_2 = 2m\pi$. And η enhanced periodically at resonance wavelength that meets this condition. The spacing of the resonant wavelength is given by the Free Spectral Range of the cavity.

The peak value of η at resonant wavelength is given as:

$$\eta = (1-R_1)(1-e^{-\alpha d})[\frac{(1+R_2e^{-\alpha d})}{(1-\sqrt{R_1R_2}e^{-\alpha_c d})^2}]$$

for a thin active layer as $\alpha d<<1$, η becomes:

$$\eta = (1-R_1)\alpha d[\frac{(1+R_2e^{-\alpha d})}{(1-\sqrt{R_1R_2}e^{-\alpha_c d})^2}]$$

This is a significant improve from the quantum efficiency of a conventional photodetector which is given by:

$$\eta = (1-R)\alpha L.$$

This shows that higher quantum efficiency can be achieved for smaller absorption region.

The critical design requirements are : a very high back mirror reflectivity and a moderate absorption layer thickness. At optical frequencies metal mirrors have low reflectivity (94%) when used on materials like GaAs. This makes metal mirrors inefficient for RCE detection. Whereas distributed Bragg reflector (DBR) can provide reflectivity near unity, are ideal choices for RCE structures.

For a R2=0.99 and $\alpha=10^4$ cm-1 with a R1=0.2 a η of 0.99 or more can be achievable for d=0.7–0.95 μm. Similarly for different values of R1 very high η is possible to achieve. However, R1 =0 limits the length of thickness region, d>5 μm can achieve 0.99 η but at the cost of bandwidth.

Detection Speed of RCE Photodiodes

The detection speed depends upon the drift velocities of the electrons and holes. And between these two holes have slower drift velocity than the electrons. The transit time limited bandwidth of conventional p-i-n photodiode is given by:

$$f_{transit} = 0.45\frac{v_h}{L}$$

However the quantum efficiency is a function of L as:

$$\eta = (1-R)\alpha L .$$

For a high speed detector for a small value of L, as α is very small, η becomes very small (η<<1). This shows for an optimum value of quantum efficiency the bandwidth has to sacrifice.

A p-i-n RCE photodetector can reduce the absorption region to a much smaller scale. In this case the carriers need to traverse a smaller distance as well, L_1 (< L) and L_2 (< L) for electrons and holes respectively.

The length of L1 and L2 can also be optimized to match the delay between the hole and electron drift. And the transition bandwidth becomes:

$$f_{transit} = 0.45\frac{v_h + v_e}{L+d}$$

As in most of semiconductors v_e is more than v_h the bandwidth increases drastically.

It's been reported that for a large device of L=0.5μm 64 GHz of bandwidth can be achieved and a small device of L=0.25μm can give 120 GHz bandwidth, where conventional photodetectors have bandwidth of 10–30 GHz.

Wavelength Selectivity of RCE Photo Detectors

A RCE structure can make the detector wavelength selective to an extent due to the resonance properties of the cavity. The resonance condition of the cavity is given as $2\beta L + \phi_1 + \phi_2 = 2m\pi$. For any other value the efficiency η reduces from its maximum value, and vanishes when $2\beta L + \phi_1 + \phi_2 = (2m+1)\pi$. The wavelength spacing of the maxima of η are separated by the Free Spectral Range of the cavity, given as:

$$FSR = \frac{\lambda^2}{2n_{eff}(L + L_{eff,1} + L_{eff,2})}$$

Where neff is the effective refractive index and Leff, i are the effective optical path lengths of the mirrors.

Finesse, the ratio of the FSR to the FWHM at the resonant wavelength, gives the wavelength selectivity of the cavity.

This shows that the wavelength selectivity increases with higher reflectivity and smaller values of L.

Material Requirements for RCE Devices

The estimated superior performance of the RCE devices critically depends on the realization of very low loss active region. This enforces the conditions that: the mirror and the cavity materials must be non-absorbing at the detection wavelength; and the mirror should have very high reflectivity so that it gives highest optical confinement inside the cavity.

The absorption in the cavity can be limited by making the bandgap of the active region smaller than the cavity and the mirror. But a large difference in the bandgap would be a blockage in extraction of photo generated carriers from a heterojunction. Usually a moderate offset is kept within the absorption spectrum.

Different material combinations satisfy all of the above criteria and are therefore usable to the RCE scheme. Some material combinations used for RCE detection are:

1. GaAs(M,C) / AlGaAs(M) / InGaAs(A) near 830-920nm.
2. InP(C) / $In_{0.53}Ga_{0.47}As$(M) / $In_{0.52}Al_{0.48}As$(M) / $In_{0.53-0.7}GaAs$(A) near 1550nm.
3. GaAs(M,C) / AlAs(M) / Ge(A) near 830-920nm.
4. Si(M,C) / SiGe(M) / Ge(A) near 1550nm.
5. GaP(M) / AlP(M) / Si(A,S) near visible region.

Future of RCE Photodiodes

There are many examples of RCE devices, like p-i-n photodiode, avalanche photodiode, schottky diode are made that verifies the theory successfully. Some of them are in use in practical purposes as well as there is a future prospect in use as modulators, optical logics in wavelength division multiplexing (WDM) systems which could enhance the quantum efficiency, operating bandwidth, wavelength selectivity.

RCE detectors are preferable in potential price and performance in commercial WDM systems. RCE detectors have very good potential for implementations in WDM systems and improve the performance significantly. There are various implementations of RCE modulators are made and there is a huge scope for further improvement in performance of those. Other than the photodetectors the RCE structures have lots of other implementations and a very high potential for improved performance. A Light Emitting Diode (LED) can be made to have narrower spectrum and higher directivity to allow more coupling to optical fibre and better utilization of the fibre bandwidth. Optical amplifiers can be made to have more compact, thus lower power required to pump and also at lower cost. Photonic logics will work more efficiently than they do. There will be much less crosstalk, more speed, more gain with simple design.

References

- Schacter, Daniel L. (2011). Psychology Second Edition. 41 Madison Avenue, New York, NY 10010: Worth Publishers. pp. 136–137. ISBN 978-1-4292-3719-2.
- Kandel, E. R.; Schwartz, J.H.; Jessell, T.M. (2000). Principles of Neural Science (4th ed.). New York: McGraw-Hill. pp. 507–513. ISBN 0-8385-7701-6.
- Enss, Christian (Editor) (2005). Cryogenic Particle Detection. Springer, Topics in applied physics 99. ISBN 3-540-20113-0.
- F. Marsili et al., "Single-photon detectors based on ultranarrow superconducting nanowires," Nano Letters 11, 2048 (2011), doi:10.1021/nl2005143, arXiv:1012.4149
- S. N. Dorenbos et al., "Low gap superconducting single photon detectors for infrared sensitivity," Applied Physics Letters 98, 251102 (2011), doi:10.1063/1.3599712
- Louvi, A.; Grove, E. A. (2011). "Cilia in the CNS: The quiet organelle claims center stage". Neuron. 69: 1046–1060. doi:10.1016/j.neuron.2011.03.002.
- Mounting Brackets & Components for Photoelectric/Automation Sensors, 2010, SoftNoze USA Inc, Frankfort, NY USA
- M. Rosticher et al., "A high efficiency superconducting nanowire single electron detector," Applied Physics Letters 97, 183106 (2010), doi:10.1063/1.3506692

6

Applications of Photodetectors

There are many allied fields of photodetection technologies that use the basic concepts of photodetectors. These are technology whose functioning and process depends on the activation of light. Some common applications of photodetectors that are listed in this chapter are active pixel sensor, solar panels and solar cells.

Active Pixel Sensor

An active-pixel sensor (APS) is an image sensor consisting of an integrated circuit containing an array of pixel sensors, each pixel containing a photodetector and an active amplifier. There are many types of active pixel sensors including the CMOS APS used most commonly in cell phone cameras, web cameras, most digital pocket cameras since 2010, and in most DSLRs. Such an image sensor is produced using CMOS technology (and is hence also known as a CMOS sensor), and has emerged as an alternative to charge-coupled device (CCD) image sensors.

CMOS image sensor

The term *active pixel sensor* is also used to refer to the individual pixel sensor itself, as opposed to the image sensor; in that case the image sensor is sometimes called an *active pixel sensor imager*, or *active-pixel image sensor*.

History

The term *active pixel sensor* was coined in 1985 by Tsutomu Nakamura who worked on the Charge Modulation Device active pixel sensor at Olympus, and more broadly defined by Eric Fossum in a 1993 paper.

Image sensor elements with in-pixel amplifiers were described by Noble in 1968, by Chamberlain in 1969, and by Weimer *et al.* in 1969, at a time when *passive-pixel sensors* – that is, pixel sensors without their own amplifiers – were being investigated as a solid-state alternative to vacuum-tube imaging devices. The MOS passive-pixel sensor used just a simple switch in the pixel to read out the photodiode integrated charge. Pixels were arrayed in a two-dimensional structure, with an access enable wire shared by pixels in the same row, and output wire shared by column. At the end of each column was an amplifier. Passive-pixel sensors suffered from many limitations, such as high noise, slow readout, and lack of scalability. The addition of an amplifier to each pixel addressed these problems, and resulted in the creation of the active-pixel sensor. Noble in 1968 and Chamberlain in 1969 created sensor arrays with active MOS readout amplifiers per pixel, in essentially the modern three-transistor configuration. The CCD was invented in October 1969 at Bell Labs. Because the MOS process was so variable and MOS transistors had characteristics that changed over time (Vth instability), the CCD's charge-domain operation was more manufacturable and quickly eclipsed MOS passive and active pixel sensors. A low-resolution "mostly digital" N-channel MOSFET imager with intra-pixel amplification, for an optical mouse application, was demonstrated in 1981.

Another type of active pixel sensor is the hybrid infrared focal plane array (IRFPA) designed to operate at cryogenic temperatures in the infrared spectrum. The devices are two chips that are put together like a sandwich: one chip contains detector elements made in InGaAs or HgCdTe, and the other chip is typically made of silicon and is used to read out the photodetectors. The exact date of origin of these devices is classified, but by the mid-1980s they were in widespread use.

By the late 1980s and early 1990s, the CMOS process was well established as a well controlled stable process and was the baseline process for almost all logic and microprocessors. There was a resurgence in the use of passive-pixel sensors for low-end imaging applications, and active-pixel sensors for low-resolution high-function applications such as retina simulation and high energy particle detector. However, CCDs continued to have much lower temporal noise and fixed-pattern noise and were the dominant technology for consumer applications such as camcorders as well as for broadcast cameras, where they were displacing video camera tubes.

Eric Fossum, *et al.*, invented the image sensor that used intra-pixel charge transfer along with an in-pixel amplifier to achieve true correlated double sampling (CDS) and low temporal noise operation, and on-chip circuits for fixed-pattern noise reduction, and published the first extensive article predicting the emergence of APS imagers as the commercial successor of CCDs. Between 1993 and 1995, the Jet Propulsion Laboratory developed a number of prototype devices, which validated the key features of the technology. Though primitive, these devices demonstrated good image performance with high readout speed and low power consumption.

In 1995, personnel from JPL founded Photobit Corp., who continued to develop and commercialize APS technology for a number of applications, such as web cams, high speed and motion capture cameras, digital radiography, endoscopy (pill) cameras, DSLRs and camera-phones. Many other small image sensor companies also sprang to life shortly thereafter due to the accessibility of the CMOS process and all quickly adopted the active pixel sensor approach. Most recent, the CMOS sensor technology has spread to medium-format photography with Phase One being the first to launch a medium format digital back with a Sony-built CMOS sensor.

Comparison to CCDs

APS pixels solve the speed and scalability issues of the passive-pixel sensor. They generally consume less power than CCDs, have less image lag, and require less specialized manufacturing facilities. Unlike CCDs, APS sensors can combine the image sensor function and image processing functions within the same integrated circuit. APS sensors have found markets in many consumer applications, especially camera phones. They have also been used in other fields including digital radiography, military ultra high speed image acquisition, security cameras, and optical mice. Manufacturers include Aptina Imaging (independent spinout from Micron Technology, who purchased Photobit in 2001), Canon, Samsung, STMicroelectronics, Toshiba, OmniVision Technologies, Sony, and Foveon, among others. CMOS-type APS sensors are typically suited to applications in which packaging, power management, and on-chip processing are important. CMOS type sensors are widely used, from high-end digital photography down to mobile-phone cameras.

Advantages of CMOS Compared to CCD

Blooming in a CCD image

A big advantage of a CMOS sensor is that it is typically less expensive than a CCD sensor.

A CMOS sensor also typically has better control of blooming (that is, of bleeding of photo-charge from an over-exposed pixel into other nearby pixels).

Disadvantages of CMOS Compared to CCD

Distortion caused by a rolling shutter

Since a CMOS sensor typically captures a row at a time within approximately 1/60th or 1/50th of a second (depending on refresh rate) it may result in a "rolling shutter" effect, where the image is skewed (tilted to the left or right, depending on the direction of camera or subject movement). For example, when tracking a car moving at high speed, the car will not be distorted but the background will appear to be tilted. A frame-transfer CCD sensor does not have this problem, instead capturing the entire image at once into a frame store.

The active circuitry in CMOS pixels takes some area on the surface which is not light-sensitive, reducing the quantum efficiency of the device. Thus, CCDs are preferred in astronomical applications.

Architecture

Pixel

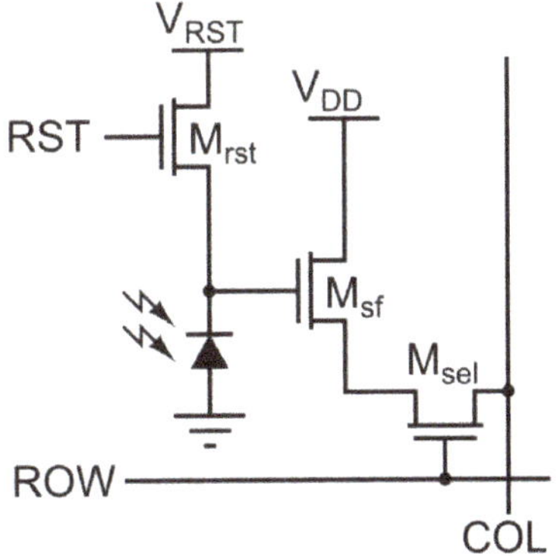

A three-transistor active pixel sensor.

The standard CMOS APS pixel today consists of a photodetector (a *pinned* photodiode), a floating diffusion, a transfer gate, reset gate, selection gate and source-follower readout transistor—the so-called 4T cell. The pinned photodiode was originally used in interline transfer CCDs due to its low dark current and good blue response, and when coupled with the transfer gate, allows complete charge transfer from the pinned photodiode to the floating diffusion (which is further connected to the gate of the read-out transistor) eliminating lag. The use of intrapixel charge transfer can offer lower noise by enabling the use of correlated double sampling (CDS). The Noble 3T pixel is still sometimes used since the fabrication requirements are easier. The 3T pixel comprises the same elements as the 4T pixel except the transfer gate and the photodiode. The reset transistor, M_{rst}, acts as a switch to reset the floating diffusion which acts in this case as the photodiode. When the reset transistor is turned on, the photodiode is effectively connected to the power supply, V_{RST}, clearing all integrated charge. Since the reset transistor is n-type, the pixel operates in soft reset. The read-out transistor, M_{sf}, acts as a buffer (specifically, a source follower), an amplifier which allows the pixel voltage to be observed without removing the accumulated charge. Its power supply, V_{DD}, is typically tied to the power supply of the reset transistor. The select transistor, M_{sel}, allows a single row of the pixel array to be read by the read-out electronics. Other innovations of the pixels such as 5T and 6T pixels also exist. By adding extra transistors, functions such as global shutter, as opposed to the more common rolling shutter, are possible. In order to increase the pixel densities, shared-row, four-ways and eight-ways shared read out, and other architectures can be employed. A variant of the 3T active pixel is the Foveon X3 sensor invented by Dick Merrill. In this device, three photodiodes are stacked on top of each other using planar fabrication techniques, each photodiode having its own 3T circuit. Each successive layer acts as a filter for the layer below it shifting the spectrum of absorbed light in successive layers. By deconvolving the response of each layered detector, red, green, and blue signals can be reconstructed.

APS Using TFTs

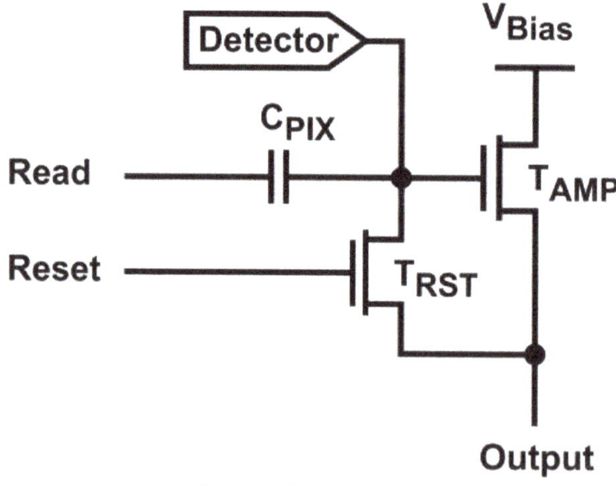

A two-transistor active/passive pixel sensor

For applications such as large-area digital X-ray imaging, thin-film transistors (TFTs) can also be used in APS architecture. However, because of the larger size and lower transconductance gain of TFTs compared to CMOS transistors, it is necessary to have fewer on-pixel TFTs to maintain image resolution and quality at an acceptable level. A two-transistor APS/PPS architecture has been shown to be promising for APS using amorphous silicon TFTs. In the two-transistor APS architecture on the right, T_{AMP} is used as a switched-amplifier integrating functions of both M_{sf} and M_{sel} in the three-transistor APS. This results in reduced transistor counts per pixel, as well as increased pixel transconductance gain. Here, C_{pix} is the pixel storage capacitance, and it is also used to capacitively couple the addressing pulse of the "Read" to the gate of T_{AMP} for ON-OFF switching. Such pixel readout circuits work best with low capacitance photoconductor detectors such as amorphous selenium.

Array

A typical two-dimensional array of pixels is organized into rows and columns. Pixels in a given row share reset lines, so that a whole row is reset at a time. The row select lines of each pixel in a row are tied together as well. The outputs of each pixel in any given column are tied together. Since only one row is selected at a given time, no competition for the output line occurs. Further amplifier circuitry is typically on a column basis.

Size

The size of the pixel sensor is often given in height and width, but also in the optical format.

Design Variants

Many different pixel designs have been proposed and fabricated. The standard pixel is the most common because it uses the fewest wires and the fewest, most tightly packed transistors possible for an active pixel. It is important that the active circuitry in a pixel take up as little space as possible to allow more room for the photodetector. High transistor count hurts fill factor, that is, the percentage of the pixel area that is sensitive to light. Pixel size can be traded for desirable qualities such as noise reduction or reduced image lag. Noise is a measure of the accuracy with which the incident light can be measured. Lag occurs when traces of a previous frame remain in future frames, i.e. the pixel is not fully reset. The voltage noise variance in a soft-reset (gate-voltage regulated) pixel is $N_e = \frac{\sqrt{kTC/2}}{q}$, but image lag and fixed pattern noise may be problematic. In rms electrons, the noise is $N_e = \frac{\sqrt{kTC/2}}{q}$.

Hard Reset

Operating the pixel via hard reset results in a Johnson–Nyquist noise on the photodiode of $V_n^2 = kT / C$ or $N_e = \frac{\sqrt{kTC}}{q}$, but prevents image lag, sometimes a desirable tradeoff. One way to use hard reset is replace M_{rst} with a p-type transistor and invert the polarity of the RST signal. The presence of the p-type device reduces fill factor, as extra space is required between p- and n-devices; it also removes the possibility of using the reset transistor as an overflow anti-blooming drain, which is a commonly exploited benefit of the n-type reset FET. Another way to achieve hard reset, with the n-type FET, is to lower the voltage of V_{RST} relative to the on-voltage of RST. This reduction may reduce headroom, or full-well charge capacity, but does not affect fill factor, unless V_{DD} is then routed on a separate wire with its original voltage.

Combinations of Hard and Soft Reset

Techniques such as flushed reset, pseudo-flash reset, and hard-to-soft reset combine soft and hard reset. The details of these methods differ, but the basic idea is the same. First, a hard reset is done, eliminating image lag. Next, a soft reset is done, causing a low noise reset without adding any lag. Pseudo-flash reset requires separating V_{RST} from V_{DD}, while the other two techniques add more complicated column circuitry. Specifically, pseudo-flash reset and hard-to-soft reset both add transistors between the pixel power supplies and the actual V_{DD}. The result is lower headroom, without affecting fill factor.

Active Reset

A more radical pixel design is the active-reset pixel. Active reset can result in much lower noise levels. The tradeoff is a complicated reset scheme, as well as either a much larger pixel or extra column-level circuitry.

Charge-Coupled Device

A specially developed CCD used for ultraviolet imaging in a wire-bonded package

A charge-coupled device (CCD) is a device for the movement of electrical charge, usually from within the device to an area where the charge can be manipulated, for example conversion into a digital value. This is achieved by "shifting" the signals between stages within the device one at a time. CCDs move charge between capacitive *bins* in the device, with the shift allowing for the transfer of charge between bins.

The CCD is a major piece of technology in digital imaging. In a CCD image sensor, pixels are represented by p-doped MOS capacitors. These capacitors are biased above the threshold for inversion when image acquisition begins, allowing the conversion of incoming photons into electron charges at the semiconductor-oxide interface; the CCD is then used to read out these charges. Although CCDs are not the only technology to allow for light detection, CCD image sensors are widely used in professional, medical, and scientific applications where high-quality image data is required. In applications with less exacting quality demands, such as consumer and professional digital cameras, active pixel sensors (CMOS) are generally used; the large quality advantage CCDs enjoyed early on has narrowed over time.

History

George E. Smith and Willard Boyle, 2009

The charge-coupled device was invented in 1969 at AT&T Bell Labs by Willard Boyle and George E. Smith. The lab was working on semiconductor bubble memory when Boyle and Smith conceived of the design of what they termed, in their notebook, "Charge 'Bubble' Devices". The device could be used as a shift register. The essence of the design was the ability to transfer charge along the surface of a semiconductor from one storage capacitor to the next. The concept was similar in principle to the bucket-brigade device (BBD), which was developed at Philips Research Labs during the late 1960s. The first patent (U.S. Patent 4,085,456) on the application of CCDs to imaging was assigned to Michael Tompsett.

The initial paper describing the concept listed possible uses as a memory, a delay line, and an imaging device. The first experimental device demonstrating the principle was a row of closely spaced metal squares on an oxidized silicon surface electrically accessed by wire bonds.

The first working CCD made with integrated circuit technology was a simple 8-bit shift register. This device had input and output circuits and was used to demonstrate its use as a shift register and as a crude eight pixel linear imaging device. Development of the device progressed at a rapid rate. By 1971, Bell researchers led by Michael Tompsett were able to capture images with simple linear devices. Several companies, including Fairchild Semiconductor, RCA and Texas Instruments, picked up on the invention and began development programs. Fairchild's effort, led by ex-Bell researcher Gil Amelio, was the first with commercial devices, and by 1974 had a linear 500-element device and a 2-D 100 x 100 pixel device. Steven Sasson, an electrical engineer working for Kodak, invented the first digital still camera using a Fairchild 100 x 100 CCD in 1975. The first KH-11 KENNAN reconnaissance satellite equipped with charge-coupled device array (800 x 800 pixels) technology for imaging was launched in December 1976. Under the leadership of Kazuo Iwama, Sony also started a large development effort on CCDs involving a significant investment. Eventually, Sony managed to mass-produce CCDs for their camcorders. Before this happened, Iwama died in August 1982; subsequently, a CCD chip was placed on his tombstone to acknowledge his contribution.

In January 2006, Boyle and Smith were awarded the National Academy of Engineering Charles Stark Draper Prize, and in 2009 they were awarded the Nobel Prize for Physics, for their invention of the CCD concept. Michael Tompsett was awarded the 2010 National Medal of Technology and Innovation for pioneering work and electronic technologies including the design and development of the first charge coupled device (CCD) imagers. He was also awarded the 2012 IEEE Edison Medal "For pioneering contributions to imaging devices including CCD Imagers, cameras and thermal imagers".

Basics of Operation

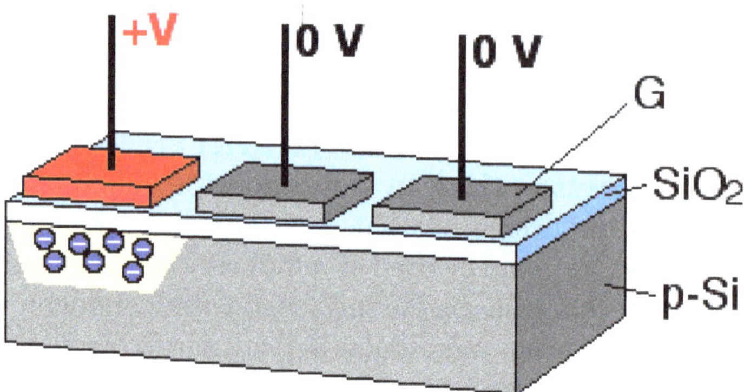

The charge packets (electrons, blue) are collected in *potential wells* (yellow) created by applying positive voltage at the gate electrodes (G). Applying positive voltage to the gate electrode in the correct sequence transfers the charge packets.

In a CCD for capturing images, there is a photoactive region (an epitaxial layer of silicon), and a transmission region made out of a shift register (the CCD, properly speaking).

An image is projected through a lens onto the capacitor array (the photoactive region), causing each capacitor to accumulate an electric charge proportional to the light intensity at that location. A one-dimensional array, used in line-scan cameras, captures a single slice of the image, whereas a two-dimensional array, used in video and still cameras, captures a two-dimensional picture corresponding to the scene projected onto the focal plane of the sensor. Once the array has been exposed to the image, a control circuit causes each capacitor to transfer its contents to its neighbor (operating as a shift register). The last capacitor in the array dumps its charge into a charge amplifier, which converts the charge into a voltage. By repeating this process, the controlling circuit converts the entire contents of the array in the semiconductor to a sequence of voltages. In a digital device, these voltages are then sampled, digitized, and usually stored in memory; in an analog device (such as an analog video camera), they are processed into a continuous analog signal (e.g. by feeding the output of the charge amplifier into a low-pass filter), which is then processed and fed out to other circuits for transmission, recording, or other processing.

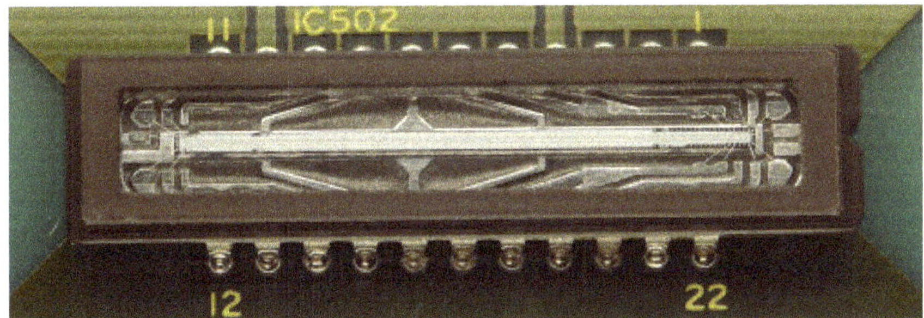

"One-dimensional" CCD image sensor from a fax machine

Detailed Physics of Operation

Charge Generation

Before the MOS capacitors are exposed to light, they are biased into the depletion region; in n-channel CCDs, the silicon under the bias gate is slightly *p*-doped or intrinsic. The gate is then biased at a positive potential, above the threshold for strong inversion, which will eventually result in the creation of a *n* channel below the gate as in a MOSFET. However, it takes time to reach this thermal equilibrium: up to hours in high-end scientific cameras cooled at low temperature. Initially after biasing, the holes are pushed far into the substrate, and no mobile electrons are at or near the surface; the CCD thus operates in a non-equilibrium state called deep depletion. Then, when electron–hole pairs are generated in the depletion region, they are separated by the electric field, the electrons move toward the surface, and the holes move toward the substrate. Four pair-generation processes can be identified:

- photo-generation (up to 95% of quantum efficiency),
- generation in the depletion region,

- generation at the surface, and
- generation in the neutral bulk.

The last three processes are known as dark-current generation, and add noise to the image; they can limit the total usable integration time. The accumulation of electrons at or near the surface can proceed either until image integration is over and charge begins to be transferred, or thermal equilibrium is reached. In this case, the well is said to be full. The maximum capacity of each well is known as the well depth, typically about 10^5 electrons per pixel.

Design and Manufacturing

The photoactive region of a CCD is, generally, an epitaxial layer of silicon. It is lightly *p* doped (usually with boron) and is grown upon a substrate material, often p++. In buried-channel devices, the type of design utilized in most modern CCDs, certain areas of the surface of the silicon are ion implanted with phosphorus, giving them an n-doped designation. This region defines the channel in which the photogenerated charge packets will travel. Simon Sze details the advantages of a buried-channel device:

This thin layer (= 0.2–0.3 micron) is fully depleted and the accumulated photogenerated charge is kept away from the surface. This structure has the advantages of higher transfer efficiency and lower dark current, from reduced surface recombination. The penalty is smaller charge capacity, by a factor of 2–3 compared to the surface-channel CCD.

The gate oxide, i.e. the capacitor dielectric, is grown on top of the epitaxial layer and substrate.

Later in the process, polysilicon gates are deposited by chemical vapor deposition, patterned with photolithography, and etched in such a way that the separately phased gates lie perpendicular to the channels. The channels are further defined by utilization of the LOCOS process to produce the channel stop region.

Channel stops are thermally grown oxides that serve to isolate the charge packets in one column from those in another. These channel stops are produced before the polysilicon gates are, as the LOCOS process utilizes a high-temperature step that would destroy the gate material. The channel stops are parallel to, and exclusive of, the channel, or "charge carrying", regions.

Channel stops often have a p+ doped region underlying them, providing a further barrier to the electrons in the charge packets (this discussion of the physics of CCD devices assumes an electron transfer device, though hole transfer is possible).

The clocking of the gates, alternately high and low, will forward and reverse bias the diode that is provided by the buried channel (n-doped) and the epitaxial layer (p-doped).

This will cause the CCD to deplete, near the p–n junction and will collect and move the charge packets beneath the gates—and within the channels—of the device.

CCD manufacturing and operation can be optimized for different uses. The above process describes a frame transfer CCD. While CCDs may be manufactured on a heavily doped p++ wafer it is also possible to manufacture a device inside p-wells that have been placed on an n-wafer. This second method, reportedly, reduces smear, dark current, and infrared and red response. This method of manufacture is used in the construction of interline-transfer devices.

Another version of CCD is called a peristaltic CCD. In a peristaltic charge-coupled device, the charge-packet transfer operation is analogous to the peristaltic contraction and dilation of the digestive system. The peristaltic CCD has an additional implant that keeps the charge away from the silicon/silicon dioxide interface and generates a large lateral electric field from one gate to the next. This provides an additional driving force to aid in transfer of the charge packets.

Architecture

The CCD image sensors can be implemented in several different architectures. The most common are full-frame, frame-transfer, and interline. The distinguishing characteristic of each of these architectures is their approach to the problem of shuttering.

In a full-frame device, all of the image area is active, and there is no electronic shutter. A mechanical shutter must be added to this type of sensor or the image smears as the device is clocked or read out.

With a frame-transfer CCD, half of the silicon area is covered by an opaque mask (typically aluminum). The image can be quickly transferred from the image area to the opaque area or storage region with acceptable smear of a few percent. That image can then be read out slowly from the storage region while a new image is integrating or exposing in the active area. Frame-transfer devices typically do not require a mechanical shutter and were a common architecture for early solid-state broadcast cameras. The downside to the frame-transfer architecture is that it requires twice the silicon real estate of an equivalent full-frame device; hence, it costs roughly twice as much.

The interline architecture extends this concept one step further and masks every other column of the image sensor for storage. In this device, only one pixel shift has to occur to transfer from image area to storage area; thus, shutter times can be less than a microsecond and smear is essentially eliminated. The advantage is not free, however, as the imaging area is now covered by opaque strips dropping the fill factor to approximately 50 percent and the effective quantum efficiency by an equivalent amount. Modern designs have addressed this deleterious characteristic by adding microlenses on the surface of the device to direct light away from the opaque regions and on the active area. Microlenses can bring the fill factor back up to 90 percent or more depending on pixel size and the overall system's optical design.

CCD from a 2.1 megapixel Argus digital camera

CCD Sony ICX493AQA 10.14 (Gross 10.75) Mpixels APS-C 1.8" (23.98 x 16.41mm) sensor side

CCD Sony ICX493AQA 10.14 (Gross 10.75) Mpixels APS-C 1.8" (23.98 x 16.41mm) pins side

The choice of architecture comes down to one of utility. If the application cannot tolerate an expensive, failure-prone, power-intensive mechanical shutter, an interline device is the right choice. Consumer snap-shot cameras have used interline devices. On the other hand, for those applications that require the best possible light collection and issues of money, power and time are less important, the full-frame device is the right choice. Astronomers tend to prefer full-frame devices. The frame-transfer falls in between and was a common choice before the fill-factor issue of interline devices was addressed. Today, frame-transfer is usually chosen when an interline architecture is not available, such as in a back-illuminated device.

CCDs containing grids of pixels are used in digital cameras, optical scanners, and video cameras as light-sensing devices. They commonly respond to 70 percent of the incident light (meaning a quantum efficiency of about 70 percent) making them far more efficient than photographic film, which captures only about 2 percent of the incident light.

CCD from a 2.1 megapixel Hewlett-Packard digital camera

Most common types of CCDs are sensitive to near-infrared light, which allows infrared photography, night-vision devices, and zero lux (or near zero lux) video-recording/photography. For normal silicon-based detectors, the sensitivity is limited to 1.1 μm. One other consequence of their sensitivity to infrared is that infrared from remote controls often appears on CCD-based digital cameras or camcorders if they do not have infrared blockers.

Cooling reduces the array's dark current, improving the sensitivity of the CCD to low light intensities, even for ultraviolet and visible wavelengths. Professional observatories often cool their detectors with liquid nitrogen to reduce the dark current, and therefore the thermal noise, to negligible levels.

Frame Transfer CCD

A frame transfer CCD sensor

The frame transfer CCD imager was the first imaging structure proposed for CCD Imaging by Michael Tompsett at Bell Laboratories. A frame transfer CCD is a specialized CCD, often used in astronomy and some professional video cameras, designed for high exposure efficiency and correctness.

The normal functioning of a CCD, astronomical or otherwise, can be divided into two phases: exposure and readout. During the first phase, the CCD passively collects incoming photons, storing electrons in its cells. After the exposure time is passed, the cells are read out one line at a time. During the readout phase, cells are shifted down the entire area of the CCD. While they are shifted, they continue to collect light. Thus, if the shifting is not fast enough, errors can result from light that falls on a cell holding charge during the transfer. These errors are referred to as "vertical smear" and cause a strong light source to create a vertical line above and below its exact location. In addition, the CCD cannot be used to collect light while it is being read out. Unfortunately, a faster shifting requires a faster readout, and a faster readout can introduce errors in the cell charge measurement, leading to a higher noise level.

A frame transfer CCD solves both problems: it has a shielded, not light sensitive, area containing as many cells as the area exposed to light. Typically, this area is covered by a reflective material such as aluminium. When the exposure time is up, the cells are transferred very rapidly to the hidden area. Here, safe from any incoming light, cells can be read out at any speed one deems necessary to correctly measure the cells' charge. At the same time, the exposed part of the CCD is collecting light again, so no delay occurs between successive exposures.

The disadvantage of such a CCD is the higher cost: the cell area is basically doubled, and more complex control electronics are needed.

Intensified Charge-coupled Device

An intensified charge-coupled device (ICCD) is a CCD that is optically connected to an image intensifier that is mounted in front of the CCD.

An image intensifier includes three functional elements: a photocathode, a micro-channel plate (MCP) and a phosphor screen. These three elements are mounted one close behind the other in the mentioned sequence. The photons which are coming from the light source fall onto the photocathode, thereby generating photoelectrons. The photoelectrons are accelerated towards the MCP by an electrical control voltage, applied between photocathode and MCP. The electrons are multiplied inside of the MCP and thereafter accelerated towards the phosphor screen. The phosphor screen finally converts the multiplied electrons back to photons which are guided to the CCD by a fiber optic or a lens.

An image intensifier inherently includes a shutter functionality: If the control voltage between the photocathode and the MCP is reversed, the emitted photoelectrons are not accelerated towards the MCP but return to the photocathode. Thus, no electrons are multiplied and emitted by the MCP, no electrons are going to the phosphor screen and no light is emitted from the image intensifier. In this case no light falls onto the CCD, which means that the shutter is closed. The process of reversing the control voltage at the photocathode is called *gating* and therefore ICCDs are also called gateable CCD cameras.

Besides the extremely high sensitivity of ICCD cameras, which enable single photon detection, the gateability is one of the major advantages of the ICCD over the EMCCD cameras. The highest performing ICCD cameras enable shutter times as short as 200 picoseconds.

ICCD cameras are in general somewhat higher in price than EMCCD cameras because they need the expensive image intensifier. On the other hand, EMCCD cameras need a cooling system to cool the EMCCD chip down to temperatures around 170 K. This cooling system adds additional costs to the EMCCD camera and often yields heavy condensation problems in the application.

ICCDs are used in night vision devices and in various scientific applications.

Electron-multiplying CCD

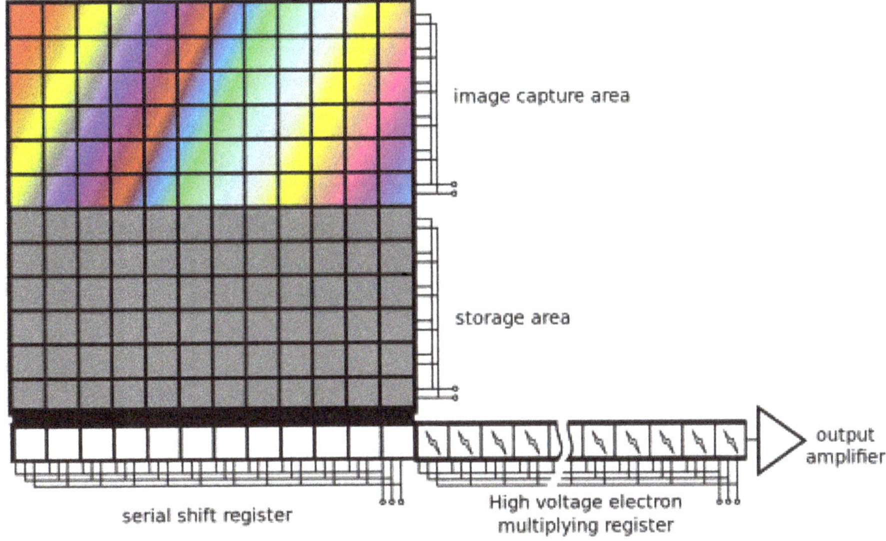

Electrons are transferred serially through the gain stages making up the multiplication register of an EMCCD. The high voltages used in these serial transfers induce the creation of additional charge carriers through impact ionisation.

An electron-multiplying **CCD** (EMCCD, also known as an L3Vision CCD, a product commercialized by e2v Ltd., GB, L3CCD or Impactron CCD, a product offered by Texas Instruments) is a charge-coupled device in which a gain register is placed between the shift register and the output amplifier. The gain register is split up into a large number of stages. In each stage, the electrons are multiplied by impact ionization in a similar way to an avalanche diode. The gain probability at every stage of the register is small ($P < 2\%$), but as the number of elements is large (N > 500), the overall gain can be very high (), with single input electrons giving many thousands of output electrons. Reading a signal from a CCD gives a noise background, typically a few electrons. In an EMCCD, this noise is superimposed on many thousands of electrons rather than a single electron; the devices' primary advantage is thus their negligible readout noise. It is to be noted that the use of avalanche breakdown for amplification of photo charges had already been described in the U.S. Patent 3,761,744 in 1973 by George E. Smith/Bell Telephone Laboratories.

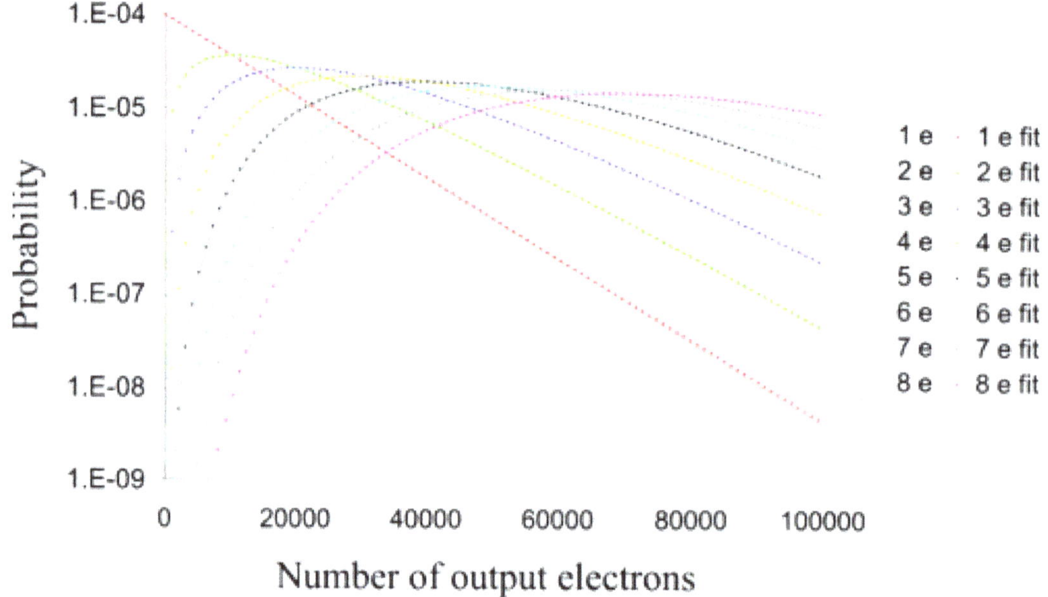

in an EMCCD there is a dispersion (variation) in the number of electrons output by the multiplication register for a given (fixed) number of input electrons. The probability distribution for the number of output electrons is plotted logarithmically on the vertical axis for a simulation of a multiplication register.

EMCCDs show a similar sensitivity to intensified CCDs (ICCDs). However, as with ICCDs, the gain that is applied in the gain register is stochastic and the *exact* gain that has been applied to a pixel's charge is impossible to know. At high gains (> 30), this uncertainty has the same effect on the signal-to-noise ratio (SNR) as halving the quantum efficiency (QE) with respect to operation with a gain of unity. However, at very low light levels (where the quantum efficiency is most important), it can be assumed that a pixel either contains an electron — or not. This removes the noise associated with the stochastic multiplication at the risk of counting multiple electrons in the same pixel as a single electron. To avoid multiple counts in one pixel due to coincident photons in this mode of operation, high frame rates are essential. The dispersion in the gain is shown in the graph on the right. For multiplication registers with many elements and large gains it is well modelled by the equation:

$$P(n)=\frac{(n-m+1)^{m-1}}{(m-1)!\left(g-1+\frac{1}{m}\right)^{m}}\exp\left(-\frac{n-m+1}{g-1+\frac{1}{m}}\right) \quad \text{if } n\geq m$$

where P is the probability of getting n output electrons given m input electrons and a total mean multiplication register gain of g.

Because of the lower costs and better resolution, EMCCDs are capable of replacing ICCDs in many applications. ICCDs still have the advantage that they can be gated very fast and thus are useful in applications like range-gated imaging. EMCCD cameras indispensably need a cooling system — using either thermoelectric cooling or liquid nitrogen — to cool the chip down to temperatures in the range of −65 to −95 °C (−85 to −139 °F). This cooling system unfortunately adds additional costs to the EMCCD imaging system and may yield condensation problems in the application. However, high-end EMCCD cameras are equipped with a permanent hermetic vacuum system confining the chip to avoid condensation issues.

The low-light capabilities of EMCCDs find use in astronomy and biomedical research, among other fields. In particular, their low noise at high readout speeds makes them very useful for a variety of astronomical applications involving low light sources and transient events such as lucky imaging of faint stars, high speed photon counting photometry, Fabry-Pérot spectroscopy and high-resolution spectroscopy. More recently, these types of CCDs have broken into the field of biomedical research in low-light applications including small animal imaging, single-molecule imaging, Raman spectroscopy, super resolution microscopy as well as a wide variety of modern fluorescence microscopy techniques thanks to greater SNR in low-light conditions in comparison with traditional CCDs and ICCDs.

In terms of noise, commercial EMCCD cameras typically have clock-induced charge (CIC) and dark current (dependent on the extent of cooling) that together lead to an effective readout noise ranging from 0.01 to 1 electrons per pixel read. However, recent improvements in EMCCD technology have led to a new generation of cameras capable of producing significantly less CIC, higher charge transfer efficiency and an EM gain 5 times higher than what was previously available. These advances in low-light detection lead to an effective total background noise of 0.001 electrons per pixel read, a noise floor unmatched by any other low-light imaging device.

Use in Astronomy

Due to the high quantum efficiencies of CCDs (for a quantum efficiency of 100%, one count equals one photon), linearity of their outputs, ease of use compared to photographic plates, and a variety of other reasons, CCDs were very rapidly adopted by astronomers for nearly all UV-to-infrared applications.

Thermal noise and cosmic rays may alter the pixels in the CCD array. To counter such effects, astronomers take several exposures with the CCD shutter closed and opened. The average of images taken with the shutter closed is necessary to lower the random noise. Once developed, the *dark frame* average image is then subtracted from the open-shutter image to remove the dark current and other systematic defects (dead pixels, hot pixels, etc.) in the CCD.

The Hubble Space Telescope, in particular, has a highly developed series of steps ("data reduction pipeline") to convert the raw CCD data to useful images.

CCD cameras used in astrophotography often require sturdy mounts to cope with vibrations from wind and other sources, along with the tremendous weight of most imaging platforms. To take long exposures of galaxies and nebulae, many astronomers use a technique known as auto-guiding. Most autoguiders use a second CCD chip to monitor deviations during imaging. This chip can rapidly detect errors in tracking and command the mount motors to correct for them.

Array of 30 CCDs used on Sloan Digital Sky Survey telescope imaging camera, an example of "drift-scanning."

An interesting unusual astronomical application of CCDs, called *drift-scanning*, uses a CCD to make a fixed telescope behave like a tracking telescope and follow the motion of the sky. The charges in the CCD are transferred and read in a direction parallel to the motion of the sky, and at the same speed. In this way, the telescope can image a larger region of the sky than its normal field of view. The Sloan Digital Sky Survey is the most famous example of this, using the technique to produce the largest uniform survey of the sky yet accomplished.

In addition to imagers, CCDs are also used in astronomical analytical instrumentation such as spectrometers.

Color Cameras

A Bayer filter on a CCD

Sony 2/3" CCD ICX024AK 10A 494496 (816*606) pixels CCD removed from video camera Sony CCD-V88E from 1988 year with vertical stripe filter Yellow, Green and Cyan

CCD color sensor

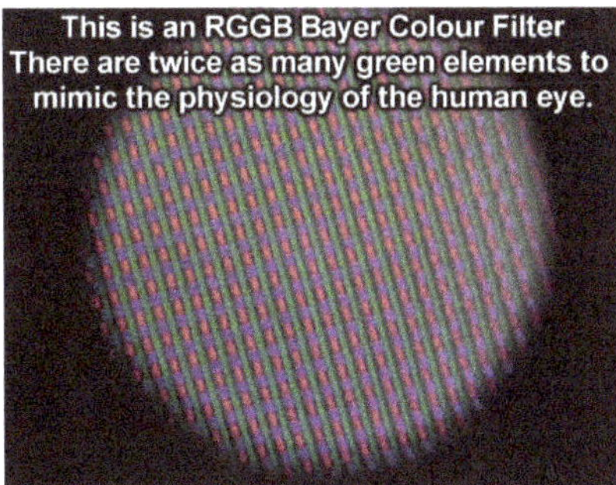

x80 microscope view of an RGGB Bayer filter on a 240 line Sony CCD PAL Camcorder CCD sensor

Digital color cameras generally use a Bayer mask over the CCD. Each square of four pixels has one filtered red, one blue, and two green (the human eye is more sensitive to green than either red or blue). The result of this is that luminance information is collected at every pixel, but the color resolution is lower than the luminance resolution.

Better color separation can be reached by three-CCD devices (3CCD) and a dichroic beam splitter prism, that splits the image into red, green and blue components. Each of the three CCDs is arranged to respond to a particular color. Many professional video camcorders, and some semi-professional camcorders, use this technique, although developments in competing CMOS technology have made CMOS sensors, both with beam-splitters and bayer filters, increasingly popular in high-end video and digital cinema cameras. Another advantage of 3CCD over a Bayer mask device is higher quantum efficiency (and therefore higher light sensitivity for a given aperture size). This is because in a 3CCD device most of the light entering the aperture is captured by a sensor, while a Bayer mask absorbs a high proportion (about 2/3) of the light falling on each CCD pixel.

For still scenes, for instance in microscopy, the resolution of a Bayer mask device can be enhanced by microscanning technology. During the process of color co-site sampling, several frames of the scene are produced. Between acquisitions, the sensor is moved in pixel dimensions, so that each point in the visual field is acquired consecutively by elements of the mask that are sensitive to the red, green and blue components of its color. Eventually every pixel in the image has been scanned at least once in each color and the resolution of the three channels become equivalent (the resolutions of red and blue channels are quadrupled while the green channel is doubled).

Sensor Sizes

Sensors (CCD / CMOS) come in various sizes, or image sensor formats. These sizes are often referred to with an inch fraction designation such as 1/1.8″ or 2/3″ called the optical format. This measurement actually originates back in the 1950s and the time of Vidicon tubes.

Blooming

Vertical smear

When a CCD exposure is long enough, eventually the electrons that collect in the "bins" in the brightest part of the image will overflow the bin, resulting in blooming. The structure of the CCD allows the electrons to flow more easily in one direction than another, resulting in vertical streaking.

Some anti-blooming features that can be built into a CCD reduce its sensitivity to light by using some of the pixel area for a drain structure. James M. Early developed a vertical anti-blooming drain that would not detract from the light collection area, and so did not reduce light sensitivity.

Solar Panel

Solar panel refers to a panel designed to absorb the sun's rays as a source of energy for generating electricity or heating.

A photovoltaic (PV) module is a packaged, connect assembly of typically 6×10 solar cells. Solar Photovoltaic panels constitute the solar array of a photovoltaic system that generates and supplies solar electricity in commercial and residential applications. Each module is rated by its DC output power under standard test conditions, and typically ranges from 100 to 365 watts. The efficiency of a module determines the area of a module given the same rated output – an 8% efficient 230 watt module will have twice the area of a 16% efficient 230 watt module. There are a few commercially available solar panels available that exceed 22% efficiency and reportedly also exceeding 24%. A single solar module can produce only a limited amount of power; most installations contain multiple modules. A photovoltaic system typically includes a panel or an array of solar modules, a solar inverter, and sometimes a battery and/or solar tracker and interconnection wiring.

The price of solar power has continued to fall so that in many countries it is cheaper than ordinary fossil fuel electricity from the grid (there is "grid parity").

Theory and Construction

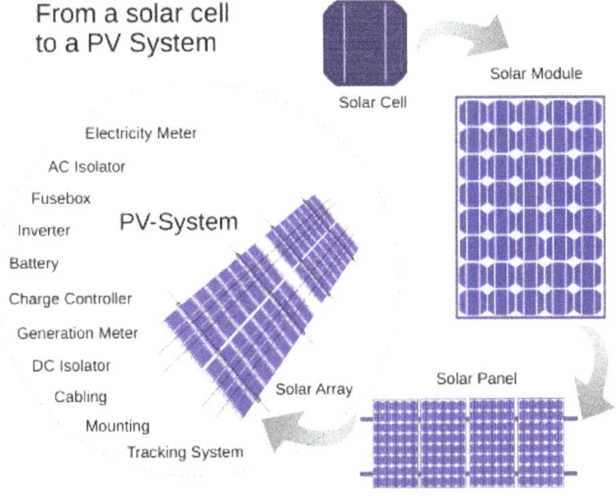

From a solar cell to a PV system

Solar modules use light energy (photons) from the sun to generate electricity through the photovoltaic effect. The majority of modules use wafer-based crystalline silicon cells or thin-film cells based on cadmium telluride or silicon. The structural (load carrying) member of a module can either be the top layer or the back layer. Cells must also be protected from mechanical damage and moisture. Most solar modules are rigid, but semi-flexible ones are available, based on thin-film cells.

Electrical connections are made in series to achieve a desired output voltage and/or in parallel to provide a desired current capability. The conducting wires that take the current off the modules may contain silver, copper or other non-magnetic conductive [transition metals]. The cells must be connected electrically to one another and to the rest of the system. Externally, popular terrestrial usage photovoltaic modules use MC3 (older) or MC4 connectors to facilitate easy weatherproof connections to the rest of the system.

Bypass diodes may be incorporated or used externally, in case of partial module shading, to maximize the output of module sections still illuminated.

Some recent solar module designs include concentrators in which light is focused by lenses or mirrors onto an array of smaller cells. This enables the use of cells with a high cost per unit area (such as gallium arsenide) in a cost-effective way.

There is also hotspot effect in solar modules in which two sides of cells are connected two diodes, this prevents solar modules or panels from short circuit.

Efficiencies

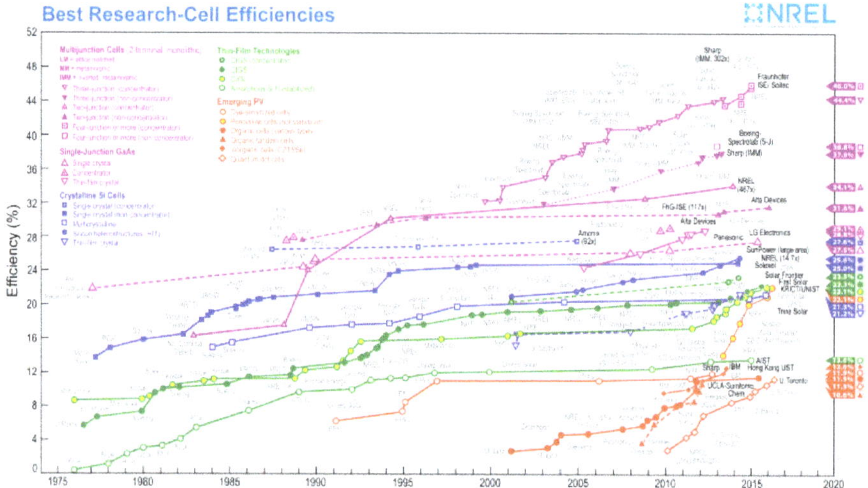

Reported timeline of solar cell energy conversion efficiencies since 1976 (National Renewable Energy Laboratory)

Depending on construction, photovoltaic modules can produce electricity from a range of frequencies of light, but usually cannot cover the entire solar range (specifically, ultraviolet, infrared and low or diffused light). Hence, much of the incident sunlight energy is wasted by solar modules, and they can give far higher efficiencies if illuminated with monochromatic light. Therefore, another design concept is to split the light into different wavelength ranges and direct the beams onto different cells tuned to those ranges. This has been projected to be capable of raising efficiency by 50%. Scientists from Spectrolab, a subsidiary of Boeing, have reported development of multi-junction solar cells with an efficiency of more than 40%, a new world record for solar photovoltaic cells. The Spectrolab scientists also predict that concentrator solar cells could achieve efficiencies of more than 45% or even 50% in the future, with theoretical efficiencies being about 58% in cells with more than three junctions.

Currently the best achieved sunlight conversion rate (solar module efficiency) is around 21.5% in new commercial products typically lower than the efficiencies of their cells in isolation. The most efficient mass-produced solar modules have power density values of up to 175 W/m^2 (16.22 W/ft^2). Research by Imperial College, London has shown that the efficiency of a solar panel can be improved by studding the light-receiving semiconductor surface with aluminum nanocylinders similar to the ridges on Lego blocks. The scattered light then travels along a longer path in the semiconductor which means that more photons can be absorbed and converted into current. Although these nanocylinders have been used previously (aluminum was preceded by gold and silver), the light scattering occurred in the near infrared region and visible light was absorbed strongly. Aluminum was found to have absorbed the ultraviolet part of the spectrum, while the visible and near infrared parts of the spectrum were found to be scattered by the aluminum surface. This, the research argued, could bring down the cost significantly and improve the efficiency as aluminum is more abundant and less costly than gold and silver. The research also noted that the increase in current makes thinner film solar panels technically feasible without "compromising power conversion efficiencies, thus reducing material consumption".

- Efficiencies of solar panel can be calculated by MPP (Maximum power point) value of solar panels
- Solar inverters convert the DC power to AC power by performing MPPT process: solar inverter samples the output Power (I-V curve) from the solar cell and applies the proper resistance (load) to solar cells to obtain maximum power.
- MPP (Maximum power point) of the solar panel consists of MPP voltage (V mpp) and MPP current (I mpp): it is a capacity of the solar panel and the higher value can make higher MPP.

Micro-inverted solar panels are wired in parallel which produces more output than normal panels which are wired in series with the output of the series determined by the lowest performing panel (this is known as the "Christmas light effect"). Micro-inverters work independently so each panel contributes its maximum possible output given the available sunlight.

Technology

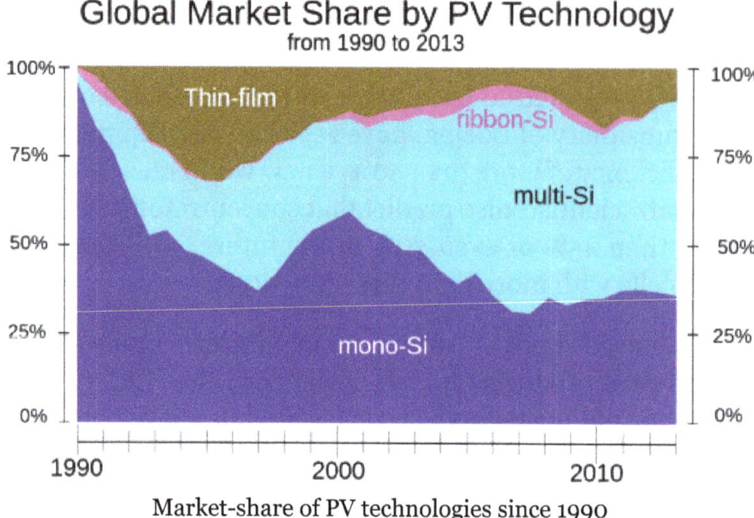

Market-share of PV technologies since 1990

Most solar modules are currently produced from crystalline silicon (c-Si) solar cells made of multicrystalline and monocrystalline silicon. In 2013, crystalline silicon accounted for more than 90 percent of worldwide PV production, while the rest of the overall market is made up of thin-film technologies using cadmium telluride, CIGS and amorphous silicon Emerging, third generation solar technologies use advanced thin-film cells. They produce a relatively high-efficiency conversion for the low cost compared to other solar technologies. Also, high-cost, high-efficiency, and close-packed rectangular multi-junction (MJ) cells are preferably used in solar panels on spacecraft, as they offer the highest ratio of generated power per kilogram lifted into space. MJ-cells are compound semiconductors and made of gallium arsenide (GaAs) and other semiconductor materials. Another emerging PV technology using MJ-cells is concentrator photovoltaics (CPV).

Thin Film

In rigid thin-film modules, the cell and the module are manufactured in the same production line. The cell is created on a glass substrate or superstrate, and the electrical connections are created *in situ*, a so-called "monolithic integration". The substrate or superstrate is laminated with an encapsulant to a front or back sheet, usually another sheet of glass. The main cell technologies in this category are CdTe, or a-Si, or a-Si+uc-Si tandem, or CIGS (or variant). Amorphous silicon has a sunlight conversion rate of 6–12%

Flexible thin film cells and modules are created on the same production line by depositing the photoactive layer and other necessary layers on a flexible substrate. If the substrate is an insulator (e.g. polyester or polyimide film) then monolithic integration can be used. If it is a conductor then another technique for electrical connection must be used. The cells are assembled into modules by laminating them to a transparent colourless fluoropolymer on the front side (typically ETFE or FEP) and a polymer suitable for bonding to the final substrate on the other side.

Smart Solar Modules

Several companies have begun embedding electronics into PV modules. This enables performing maximum power point tracking (MPPT) for each module individually, and the measurement of performance data for monitoring and fault detection at module level. Some of these solutions make use of power optimizers, a DC-to-DC converter technology developed to maximize the power harvest from solar photovoltaic systems. As of about 2010, such electronics can also compensate for shading effects, wherein a shadow falling across a section of a module causes the electrical output of one or more strings of cells in the module to fall to zero, but not having the output of the entire module fall to zero.

Performance and Degradation

Module performance is generally rated under standard test conditions (STC): irradiance of 1,000 W/m^2, solar spectrum of AM 1.5 and module temperature at 25 °C.

Electrical characteristics include nominal power (P_{MAX}, measured in W), open circuit voltage (V_{OC}), short circuit current (I_{SC}, measured in amperes), maximum power voltage (V_{MPP}), maximum power current (I_{MPP}), peak power, (watt-peak, W_p), and module efficiency (%).

Nominal voltage refers to the voltage of the battery that the module is best suited to charge; this is a leftover term from the days when solar modules were only used to charge batteries. The actual voltage output of the module changes as lighting, temperature and load conditions change, so there is never one specific voltage at which the module operates. Nominal voltage allows users, at a glance, to make sure the module is compatible with a given system.

Open circuit voltage or V_{OC} is the maximum voltage that the module can produce when not connected to an electrical circuit or system. V_{OC} can be measured with a voltmeter directly on an illuminated module's terminals or on its disconnected cable.

The peak power rating, W_p, is the maximum output under standard test conditions (not the maximum possible output). Typical modules, which could measure approximately 1x2 meters or 2x4 feet, will be rated from as low as 75 watts to as high as 350 watts, depending on their efficiency. At the time of testing, the test modules are binned according to their test results, and a typical manufacturer might rate their modules in 5 watt increments, and either rate them at +/- 3%, +/-5%, +3/-0% or +5/-0%.

Solar modules must withstand rain, hail, heavy snow load, and cycles of heat and cold for many years. Many crystalline silicon module manufacturers offer a warranty that guarantees electrical production for 10 years at 90% of rated power output and 25 years at 80%.

Potential induced degradation (also called PID) is a potential induced performance degradation in crystalline photovoltaic modules, caused by so-called stray currents. This effect may cause power loss of up to 30 percent.

The largest challenge of photovoltaic technology is the efficiencies of such solar systems. While utilizing such systems draws a great interest due to the long term returns of profit, the efficacy needs to come a long way before making it plausible to be introduced in all consumers of electricity.

The problem resides in the enormous activation energy that must be overcome for a photon to excite an electron for harvesting purposes. Advancements in photovoltaic technologies have brought about the process of "doping" the silicon substrate to lower the activation energy thereby making the panel more efficient in converting photons to retrievable electrons. Chemicals such as Boron (p-type) are applied into the semiconductor crystal in order to create donor and acceptor energy levels substantially closer to the valence and conductor bands. In doing so, the addition of Boron impurity allows the activation energy to decrease 20 fold from 1.12 eV to 0.05 eV. Since the potential difference (E_B) is so low, the Boron is able to thermally ionize at room temperatures. This allows for free energy carriers in the conduction and valence bands thereby allowing greater conversion of photons to electrons.

Solar power allows for greater efficiency than heat, such as the generation of energy in heat engines. The drawback with heat is that most of the heat created is lost to the surroundings. Thermal efficiency is as defined:

$$\eta_{th} \equiv \frac{W_{out}}{Q_{in}} = 1 - \frac{Q_{out}}{Q_{in}}$$

Due to the inherent irreversibility of heat production for useful work, efficiency levels are decreased. On the other hand, with solar panels there isn't a requirement to retain any heat, and there are no drawbacks such as friction.

Maintenance

Solar panel conversion efficiency, typically in the 20 percent range, is reduced by dust, grime, pollen, and other particulates that accumulate on the solar panel. "A dirty solar panel can reduce its power capabilities by up to 30 percent in high dust/pollen or desert areas", says Seamus Curran, associate professor of physics at the University of Houston and director of the Institute for NanoEnergy, which specializes in the design, engineering, and assembly of nanostructures.

Paying to have solar panels cleaned is often not a good investment; researchers found panels that hadn't been cleaned, or rained on, for 145 days during a summer drought in California, lost only 7.4 percent of their efficiency. Overall, for a typical residential solar system of 5 kilowatts, washing panels halfway through the summer would translate into a mere $20 gain in electricity production until the summer drought ends—in about 2 ½ months. For larger commercial rooftop systems, the financial losses are bigger but still rarely enough to warrant the cost of washing the panels. On average, panels lost a little less than 0.05 percent of their overall efficiency per day.

Recycling

Most parts of a solar module can be recycled including up to 97% of certain semiconductor materials or the glass as well as large amounts of ferrous and non-ferrous metals. Some private companies and non-profit organizations are currently engaged in take-back and recycling operations for end-of-life modules.

Recycling possibilities depend on the kind of technology used in the modules:

- Silicon based modules: aluminum frames and junction boxes are dismantled manually at the beginning of the process. The module is then crushed in a mill and the different fractions are separated - glass, plastics and metals. It is possible to recover more than 80% of the incoming weight. This process can be performed by flat glass recyclers since morphology and composition of a PV module is similar to those flat glasses used in the building and automotive industry. The recovered glass for example is readily accepted by the glass foam and glass insulation industry.
- Non-silicon based modules: they require specific recycling technologies such as the use of chemical baths in order to separate the different semiconductor materials. For cadmium telluride modules, the recycling process begins by crushing the module and subsequently separating the different fractions. This recycling process is designed to recover up to 90% of the glass and 95% of the semiconductor materials contained. Some commercial-scale recycling facilities have been created in recent years by private companies.

Since 2010, there is an annual European conference bringing together manufacturers, recyclers and researchers to look at the future of PV module recycling.

Production

#	Top Module Producer	Shipments in 2014 (MW)
1.	Yingli	3,200
2.	Trina Solar	2,580
3.	Sharp Solar	2,100
4.	Canadian Solar	1,894
5.	Jinko Solar	1,765

6.	ReneSola	1,728
7.	First Solar	1,600
8.	Hanwha SolarOne	1,280
9.	Kyocera	1,200
10.	JA Solar	1,173

In 2010, 15.9 GW of solar PV system installations were completed, with solar PV pricing survey and market research company PVinsights reporting growth of 117.8% in solar PV installation on a year-on-year basis.

With over 100% year-on-year growth in PV system installation, PV module makers dramatically increased their shipments of solar modules in 2010. They actively expanded their capacity and turned themselves into gigawatt GW players. According to PVinsights, five of the top ten PV module companies in 2010 are GW players. Suntech, First Solar, Sharp, Yingli and Trina Solar are GW producers now, and most of them doubled their shipments in 2010.

The basis of producing solar panels revolves around the use of silicon cells. These silicon cells are typically 10-20% efficient at converting sunlight into electricity, with newer production models now exceeding 22%.

In order for solar panels to become more efficient, researchers across the world have been trying to develop new technologies to make solar panels more effective at turning sunlight into energy.

In 2014, the world's top ten solar module producers in terms of shipped capacity during the calendar year of 2014 were Trina Solar, Yingli, Sharp Solar and Canadian Solar.

Price

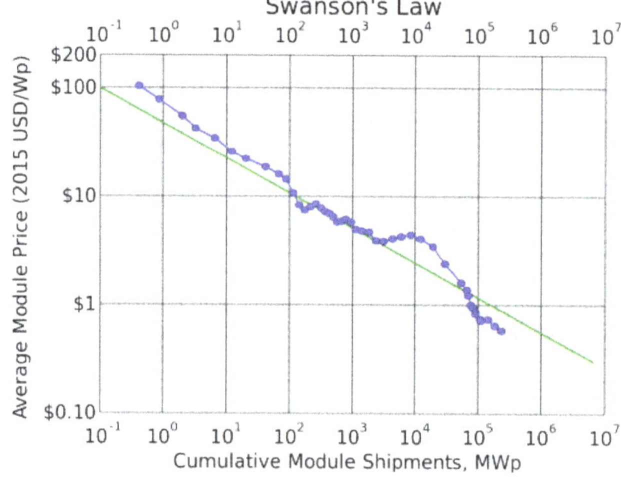

Swanson's law states that with every doubling of production of panels, there has been a 20 percent reduction in the cost of panels.

Average pricing information divides in three pricing categories: those buying small quantities (modules of all sizes in the kilowatt range annually), mid-range buyers (typically up to 10 MWp annually), and large quantity buyers (self-explanatory—and with access to the lowest prices). Over the long term there is clearly a systematic reduction in the price of cells and modules. For example, in 2012 it was estimated that the quantity cost per watt was about US$0.60, which was 250 times lower than the cost in 1970 of US$150. A 2015 study shows price/kWh dropping by 10% per year since 1980, and predicts that solar could contribute 20% of total electricity consumption by 2030, whereas the International Energy Agency predicts 16% by 2050.

Real world energy production costs depend a great deal on local weather conditions. In a cloudy country such as the United Kingdom, the cost per produced kWh is higher than in sunnier countries like Spain.

Following to RMI, Balance-of-System (BoS) elements, this is, non-module cost of non-microinverter solar modules (as wiring, converters, racking systems and various components) make up about half of the total costs of installations.

For merchant solar power stations, where the electricity is being sold into the electricity transmission network, the cost of solar energy will need to match the wholesale electricity price. This point is sometimes called 'wholesale grid parity' or 'busbar parity'.

Some photovoltaic systems, such as rooftop installations, can supply power directly to an electricity user. In these cases, the installation can be competitive when the output cost matches the price at which the user pays for his electricity consumption. This situation is sometimes called 'retail grid parity', 'socket parity' or 'dynamic grid parity'. Research carried out by UN-Energy in 2012 suggests areas of sunny countries with high electricity prices, such as Italy, Spain and Australia, and areas using diesel generators, have reached retail grid parity.

Mounting and Tracking

Solar modules mounted on solar trackers

Ground mounted photovoltaic system are usually large, utility-scale solar power plants. Their solar modules are held in place by racks or frames that are attached to ground based mounting supports. Ground based mounting supports include:

- Pole mounts, which are driven directly into the ground or embedded in concrete.
- Foundation mounts, such as concrete slabs or poured footings
- Ballasted footing mounts, such as concrete or steel bases that use weight to secure the solar module system in position and do not require ground penetration. This type of mounting system is well suited for sites where excavation is not possible such as capped landfills and simplifies decommissioning or relocation of solar module systems.

Roof-mounted solar power systems consist of solar modules held in place by racks or frames attached to roof-based mounting supports. Roof-based mounting supports include:

- Pole mounts, which are attached directly to the roof structure and may use additional rails for attaching the module racking or frames.
- Ballasted footing mounts, such as concrete or steel bases that use weight to secure the panel system in position and do not require through penetration. This mounting method allows for decommissioning or relocation of solar panel systems with no adverse effect on the roof structure.
- All wiring connecting adjacent solar modules to the energy harvesting equipment must be installed according to local electrical codes and should be run in a conduit appropriate for the climate conditions

Solar trackers increase the amount of energy produced per module at a cost of mechanical complexity and need for maintenance. They sense the direction of the Sun and tilt or rotate the modules as needed for maximum exposure to the light. Alternatively, fixed racks hold modules stationary as the sun moves across the sky. The fixed rack sets the angle at which the module is held. Tilt angles equivalent to an installation's latitude are common. Most of these fixed racks are set on poles above ground. Panels that face West or East may provide slightly lower energy, but evens out the supply, and may provide more power during peak demand.

Standards

Standards generally used in photovoltaic modules:

- IEC 61215 (crystalline silicon performance), 61646 (thin film performance) and 61730 (all modules, safety)

- ISO 9488 Solar energy—Vocabulary.
- UL 1703 From Underwriters Laboratories
- UL 1741 From Underwriters Laboratories
- UL 2703 From Underwriters Laboratories
- CE mark
- Electrical Safety Tester (EST) Series (EST-460, EST-22V, EST-22H, EST-110).

Applications

There are many practical applications for the use of solar panels or photovoltaics. It can first be used in agriculture as a power source for irrigation. In health care solar panels can be used to refrigerate medical supplies. It can also be used for infrastructure. PV modules are used in photovoltaic systems and include a large variety of electric devices:

- Photovoltaic power stations
- Rooftop solar PV systems
- Standalone PV systems
- Solar hybrid power systems
- Concentrated photovoltaics
- Solar planes
- Solar-pumped lasers
- Solar vehicles
- Solar panels on spacecrafts and space stations

Solar Cell

A solar cell, or photovoltaic cell (in very early days also termed "solar battery" – a denotation which nowadays has a totally different meaning, see here), is an electrical device that converts the energy of light directly into electricity by the photovoltaic effect, which is a physical and chemical phenomenon. It is a form of photoelectric cell, defined as a device whose electrical characteristics, such as current, voltage, or resistance, vary when exposed to light. Solar cells are the building blocks of photovoltaic modules, otherwise known as solar panels.

Solar cells are described as being photovoltaic irrespective of whether the source is sunlight or an artificial light. They are used as a photodetector (for example infrared detectors), detecting light or other electromagnetic radiation near the visible range, or measuring light intensity.

The operation of a photovoltaic (PV) cell requires 3 basic attributes:

- The absorption of light, generating either electron-hole pairs or excitons.
- The separation of charge carriers of opposite types.
- The separate extraction of those carriers to an external circuit.

In contrast, a solar thermal collector supplies heat by absorbing sunlight, for the purpose of either direct heating or indirect electrical power generation from heat. A "photoelectrolytic cell" (photoelectrochemical cell), on the other hand, refers either to a type of photovoltaic cell (like that developed by Edmond Becquerel and modern dye-sensitized solar cells), or to a device that splits water directly into hydrogen and oxygen using only solar illumination.

Applications

Assemblies of solar cells are used to make solar modules which generate electrical power from sunlight, as distinguished from a "solar thermal module" or "solar hot water panel". A solar array generates solar power using solar energy.

Cells, Modules, Panels and Systems

Typical PV system prices in 2013 in selected countries (USD)

USD/W	Australia	China	France	Germany	Italy	Japan	United Kingdom	United States
Residential	1.8	1.5	4.1	2.4	2.8	4.2	2.8	4.9
Commercial	1.7	1.4	2.7	1.8	1.9	3.6	2.4	4.5
Utility-scale	2.0	1.4	2.2	1.4	1.5	2.9	1.9	3.3

Source: *IEA – Technology Roadmap: Solar Photovoltaic Energy report*, 2014 edition Note: *DOE – Photovoltaic System Pricing Trends* reports lower prices for the U.S.

Multiple solar cells in an integrated group, all oriented in one plane, constitute a solar photovoltaic panel or solar photovoltaic module. Photovoltaic modules often have a sheet of glass on the sun-facing side, allowing light to pass while protecting the semiconductor wafers. Solar cells are usually connected in series and parallel circuits or series in modules, creating an additive voltage. Connecting cells in parallel yields a higher current; however, problems such as shadow effects can shut down the weaker (less illuminated) parallel string (a number of series connected cells) causing substantial power loss and possible damage because of the reverse bias applied to the shadowed cells by their illuminated partners. Strings of series cells are usually handled independently and not connected in parallel, though as of 2014, individual power boxes are often supplied for each module, and are connected in parallel. Although modules can be interconnected to create an array with the desired peak DC voltage and loading current capacity, using independent MPPTs (maximum power point trackers) is preferable. Otherwise, shunt diodes can reduce shadowing power loss in arrays with series/parallel connected cells.

History

The photovoltaic effect was experimentally demonstrated first by French physicist Edmond Becquerel. In 1839, at age 19, he built the world's first photovoltaic cell in his father's laboratory. Willoughby Smith first described the "Effect of Light on Selenium during the passage of an Electric Current" in a 20 February 1873 issue of Nature. In 1883 Charles Fritts built the first solid state photovoltaic cell by coating the semiconductor selenium with a thin layer of gold to form the junctions; the device was only around 1% efficient.

In 1888 Russian physicist Aleksandr Stoletov built the first cell based on the outer photoelectric effect discovered by Heinrich Hertz in 1887.

In 1905 Albert Einstein proposed a new quantum theory of light and explained the photoelectric effect in a landmark paper, for which he received the Nobel Prize in Physics in 1921.

Vadim Lashkaryov discovered *p-n*-junctions in Cu_2O and silver sulphide protocells in 1941.

Russell Ohl patented the modern junction semiconductor solar cell in 1946 while working on the series of advances that would lead to the transistor.

The first practical photovoltaic cell was publicly demonstrated on 25 April 1954 at Bell Laboratories. The inventors were Daryl Chapin, Calvin Souther Fuller and Gerald Pearson.

Solar cells gained prominence with their incorporation onto the 1958 Vanguard I satellite.

Space Applications

Solar cells were first used in a prominent application when they were proposed and flown on the Vanguard satellite in 1958, as an alternative power source to the primary battery power source. By adding cells to the outside of the body, the mission time could be extended with no major changes to the spacecraft or its power systems. In 1959 the United States launched Explorer 6, featuring large wing-shaped solar arrays, which became a common feature in satellites. These arrays consisted of 9600 Hoffman solar cells.

By the 1960s, solar cells were (and still are) the main power source for most Earth orbiting satellites and a number of probes into the solar system, since they offered the best power-to-weight ratio. However, this success was possible because in the space application, power system costs could be high, because space users had few other power options, and were willing to pay for the best possible cells. The space power market drove the development of higher efficiencies in solar cells up until the National Science Foundation "Research Applied to National Needs" program began to push development of solar cells for terrestrial applications.

In the early 1990s the technology used for space solar cells diverged from the silicon technology used for terrestrial panels, with the spacecraft application shifting to gallium arsenide-based III-V semiconductor materials, which then evolved into the modern III-V multijunction photovoltaic cell used on spacecraft.

Price Reductions

Improvements were gradual over the 1960s. This was also the reason that costs remained high, because space users were willing to pay for the best possible cells, leaving no reason to invest in lower-cost, less-efficient solutions. The price was determined largely by the semiconductor industry; their move to integrated circuits in the 1960s led to the availability of larger boules at lower relative prices. As their price fell, the price of the resulting cells did as well. These effects lowered 1971 cell costs to some $100 per watt.

In late 1969 Elliot Berman joined Exxon's task force which was looking for projects 30 years in the future and in April 1973 he founded Solar Power Corporation, a wholly owned subsidiary of Exxon at that time. The group had concluded that electrical power would be much more expensive by 2000, and felt that this increase in price would make alternative energy sources more attractive. He conducted a market study and concluded that a price per watt of about $20/watt would create significant demand. The team eliminated the steps of polishing the wafers and coating them with an anti-reflective layer, relying on the rough-sawn wafer surface. The team also replaced the expensive materials and hand wiring used in space applications with a printed circuit board on the back, acrylic plastic on the front, and silicone glue between the two, "potting" the cells. Solar cells could be made using cast-off material from the electronics market. By 1973 they announced a product, and SPC convinced Tideland Signal to use its panels to power navigational buoys, initially for the U.S. Coast Guard.

Research and Industrial Production

Research into solar power for terrestrial applications became prominent with the U.S. National Science Foundation's Advanced Solar Energy Research and Development Division within the "Research Applied to National Needs" program, which ran from 1969 to 1977, and funded research on developing solar power for ground electrical power systems. A 1973 conference, the "Cherry Hill Conference", set forth the technology goals required to achieve this goal and outlined an ambitious project for achieving them, kicking off an applied research program that would be ongoing for several decades. The program was eventually taken over by the Energy Research and Development Administration (ERDA), which was later merged into the U.S. Department of Energy.

Following the 1973 oil crisis, oil companies used their higher profits to start (or buy) solar firms, and were for decades the largest producers. Exxon, ARCO, Shell, Amoco (later purchased by BP) and Mobil all had major solar divisions during the 1970s and 1980s. Technology companies also participated, including General Electric, Motorola, IBM, Tyco and RCA.

Declining Costs and Exponential Growth

Adjusting for inflation, it cost $96 per watt for a solar module in the mid-1970s. Process improvements and a very large boost in production have brought that figure down 99%, to 68¢ per watt in 2016, according to data from Bloomberg New Energy Finance. Swanson's law is an observation similar to Moore's Law that states that solar cell prices fall 20% for every doubling of industry capacity. It was featured in an article in the British weekly newspaper The Economist.

Further improvements reduced production cost to under $1 per watt, with wholesale costs well under $2. Balance of system costs were then higher than those of the panels. Large commercial arrays could be built, as of 2010, at below $3.40 a watt, fully commissioned.

As the semiconductor industry moved to ever-larger boules, older equipment became inexpensive. Cell sizes grew as equipment became available on the surplus market; ARCO Solar's original panels used cells 2 to 4 inches (50 to 100 mm) in diameter. Panels in the 1990s and early 2000s generally used 125 mm wafers; since 2008 almost all new panels use 156 mm cells. The widespread introduction of flat screen televisions in the late 1990s and early 2000s led to the wide availability of large, high-quality glass sheets to cover the panels.

During the 1990s, polysilicon ("poly") cells became increasingly popular. These cells offer less efficiency than their monosilicon ("mono") counterparts, but they are grown in large vats that reduce cost. By the mid-2000s, poly was dominant in the low-cost panel market, but more recently the mono returned to widespread use.

Manufacturers of wafer-based cells responded to high silicon prices in 2004–2008 with rapid reductions in silicon consumption. In 2008, according to Jef Poortmans, director of IMEC's organic and solar department, current cells use 8–9 grams (0.28–0.32 oz) of silicon per watt of power generation, with wafer thicknesses in the neighborhood of 200 microns. Crystalline silicon panels dominate worldwide markets and are mostly manufactured in China and Taiwan. By late 2011, a drop in European demand due to budgetary turmoil dropped prices for crystalline solar modules to about $1.09 per watt down sharply from 2010. Prices continued to fall in 2012, reaching $0.62/watt by 4Q2012.

Global installed PV capacity reached at least 177 gigawatts in 2014, enough to supply 1 percent of the world's total electricity consumption. Solar PV is growing fastest in Asia, with China and Japan currently accounting for half of worldwide deployment.

Subsidies and grid parity

Solar-specific feed-in tariffs vary by country and within countries. Such tariffs encourage the development of solar power projects. Widespread grid parity, the point at which photovoltaic electricity is equal to or cheaper than grid power without subsidies, likely requires advances on all three fronts. Proponents of solar hope to achieve grid parity first in areas with abundant sun and high electricity costs such as in California and Japan. In 2007 BP claimed grid parity for Hawaii and other islands that otherwise use diesel fuel to produce electricity. George W. Bush set 2015 as the date for grid parity in the US. The Photovoltaic Association reported in 2012 that Australia had reached grid parity (ignoring feed in tariffs).

The price of solar panels fell steadily for 40 years, interrupted in 2004 when high subsidies in Germany drastically increased demand there and greatly increased the price of purified silicon (which is used in computer chips as well as solar panels). The recession of 2008 and the onset of Chinese manufacturing caused prices to resume their decline. In the four years after January 2008 prices for solar modules in Germany dropped from €3 to €1 per peak watt. During that same time production capacity surged with an annual growth of more than 50%. China increased market share from 8% in 2008 to over 55% in the last quarter of 2010. In December 2012 the price of Chinese solar panels had dropped to $0.60/Wp (crystalline modules).

Theory

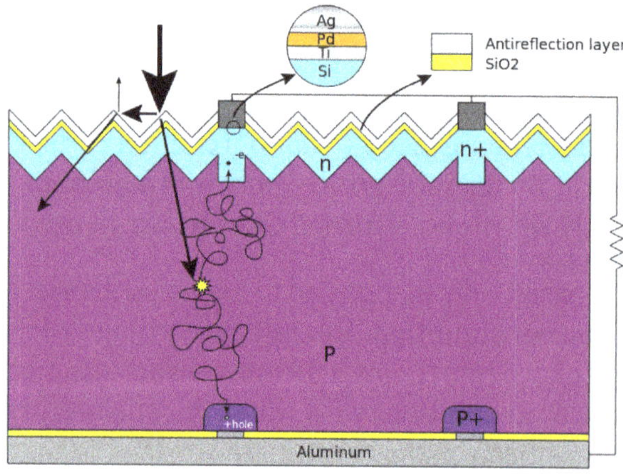

Working mechanism of a solar cell

The solar cell works in several steps:

- Photons in sunlight hit the solar panel and are absorbed by semiconducting materials, such as silicon.
- Electrons are excited from their current molecular/atomic orbital. Once excited an electron can either dissipate the energy as heat and return to its orbital or travel through the cell until it reaches an electrode. Current flows through the material to cancel the potential and this electricity is captured. The chemical bonds of the material are vital for this process to work, and usually silicon is used in two layers, one layer being bonded with boron, the other phosphorus. These layers have different chemical electric charges and subsequently both drive and direct the current of electrons.
- An array of solar cells converts solar energy into a usable amount of direct current (DC) electricity.
- An inverter can convert the power to alternating current (AC).

The most commonly known solar cell is configured as a large-area p–n junction made from silicon.

Efficiency

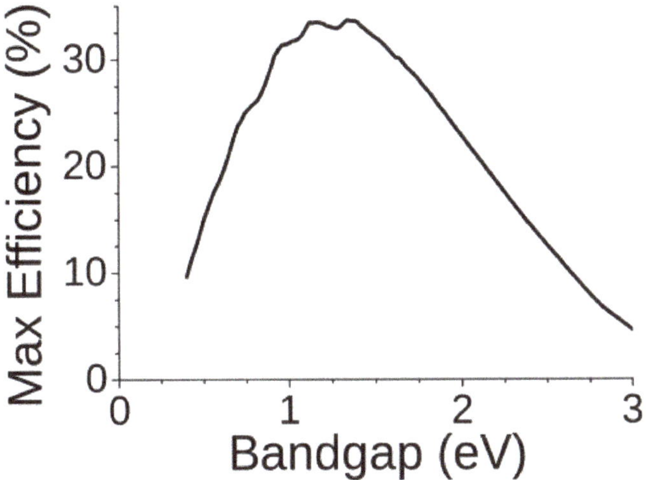

The Shockley-Queisser limit for the theoretical maximum efficiency of a solar cell. Semiconductors with band gap between 1 and 1.5eV, or near-infrared light, have the greatest potential to form an efficient single-junction cell.

Solar cell efficiency may be broken down into reflectance efficiency, thermodynamic efficiency, charge carrier separation efficiency and conductive efficiency. The overall efficiency is the product of these individual metrics.

A solar cell has a voltage dependent efficiency curve, temperature coefficients, and allowable shadow angles.

Due to the difficulty in measuring these parameters directly, other parameters are substituted: thermodynamic efficiency, quantum efficiency, integrated quantum efficiency, V_{OC} ratio, and fill factor. Reflectance losses are a portion of quantum efficiency under "external quantum efficiency". Recombination losses make up another portion of quantum efficiency, V_{OC} ratio, and fill factor. Resistive losses are predominantly categorized under fill factor, but also make up minor portions of quantum efficiency, V_{OC} ratio.

The fill factor is the ratio of the actual maximum obtainable power to the product of the open circuit voltage and short circuit current. This is a key parameter in evaluating performance. In 2009, typical commercial solar cells had a fill factor > 0.70. Grade B cells were usually between 0.4 and 0.7. Cells with a high fill factor have a low equivalent series resistance and a high equivalent shunt resistance, so less of the current produced by the cell is dissipated in internal losses.

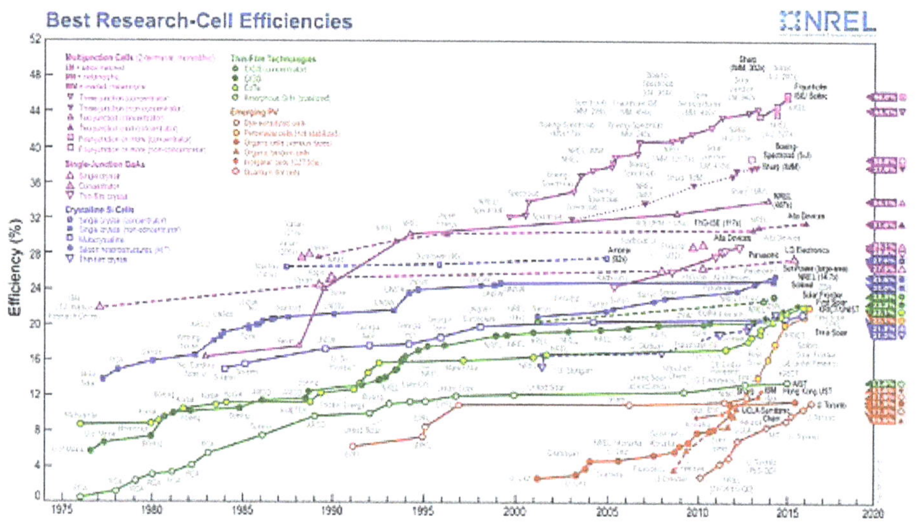

Reported timeline of solar cell energy conversion efficiencies (National Renewable Energy Laboratory)

Single p–n junction crystalline silicon devices are now approaching the theoretical limiting power efficiency of 33.7%, noted as the Shockley–Queisser limit in 1961. In the extreme, with an infinite number of layers, the corresponding limit is 86% using concentrated sunlight.

In December 2014, a solar cell achieved a new laboratory record with 46 percent efficiency in a French-German collaboration.

In 2014, three companies broke the record of 25.6% for a silicon solar cell. Panasonic's was the most efficient. The company moved the front contacts to the rear of the panel, eliminating shaded areas. In addition they applied thin silicon films to the (high quality silicon) wafer's front and back to eliminate defects at or near the wafer surface.

In September 2015, the Fraunhofer Institute for Solar Energy Systems (Fraunhofer ISE) announced the achievement of an efficiency above 20% for epitaxial wafer cells. The work on optimizing the atmospheric-pressure chemical vapor deposition (APCVD) in-line production chain was done in collaboration with NexWafe GmbH, a company spun off from Fraunhofer ISE to commercialize production.

For triple-junction thin-film solar cells, the world record is 13.6%, set in June 2015.

Materials

Solar cells are typically named after the semiconducting material they are made of. These materials must have certain characteristics in order to absorb sunlight. Some cells are designed to handle sunlight that reaches the Earth's surface, while others are optimized for use in space. Solar cells can be made of only one single layer of light-absorbing material (single-junction) or use multiple physical configurations (multi-junctions) to take advantage of various absorption and charge separation mechanisms.

Solar cells can be classified into first, second and third generation cells. The first generation cells—also called conventional, traditional or wafer-based cells—are made of crystalline silicon, the commercially predominant PV technology, that includes materials such as polysilicon and monocrystalline silicon. Second generation cells are thin film solar cells, that include amorphous silicon, CdTe and CIGS cells and are commercially significant in utility-scale photovoltaic power stations, building integrated photovoltaics or in small stand-alone power system. The third generation of solar cells includes a number of thin-film technologies often described as emerging photovoltaics—most of them have not yet been commercially applied and are still in the research or development phase. Many use organic materials, often organometallic compounds as well as inorganic substances. Despite the fact that their efficiencies had been low and the stability of the absorber material was often too short for commercial applications, there is a lot of research invested into these technologies as they promise to achieve the goal of producing low-cost, high-efficiency solar cells.

Crystalline Silicon

By far, the most prevalent bulk material for solar cells is crystalline silicon (c-Si), also known as "solar grade silicon". Bulk silicon is separated into multiple categories according to crystallinity and crystal size in the resulting ingot, ribbon or wafer. These cells are entirely based around the concept of a p-n junction. Solar cells made of c-Si are made from wafers between 160 and 240 micrometers thick.

Monocrystalline Silicon

Monocrystalline silicon (mono-Si) solar cells are more efficient and more expensive than most other types of cells. The corners of the cells look clipped, like an octagon, because the wafer material is cut from cylindrical ingots, that are typically grown by the Czochralski process. Solar panels using mono-Si cells display a distinctive pattern of small white diamonds.

Epitaxial Silicon

Epitaxial wafers can be grown on a monocrystalline silicon "seed" wafer by atmospheric-pressure CVD in a high-throughput inline process, and then detached as self-supporting wafers of some standard thickness (e.g., 250 μm) that can be manipulated by hand, and directly substituted for wafer cells cut from monocrystalline silicon ingots. Solar cells made with this technique can have efficiencies approaching those of wafer-cut cells, but at appreciably lower cost.

Polycrystalline Silicon

Polycrystalline silicon, or multicrystalline silicon (multi-Si) cells are made from cast square ingots—large blocks of molten silicon carefully cooled and solidified. They consist of small crystals giving the material its typical metal flake effect. Polysilicon cells are the most common type used in photovoltaics and are less expensive, but also less efficient, than those made from monocrystalline silicon.

Ribbon Silicon

Ribbon silicon is a type of polycrystalline silicon—it is formed by drawing flat thin films from molten silicon and results in a polycrystalline structure. These cells are cheaper to make than multi-Si, due to a great reduction in silicon waste, as this approach does not require sawing from ingots. However, they are also less efficient.

Mono-like-multi Silicon (MLM)

This form was developed in the 2000s and introduced commercially around 2009. Also called cast-mono, this design uses polycrystalline casting chambers with small "seeds" of mono material. The result is a bulk mono-like material that is polycrystalline around the outsides. When sliced for processing, the inner sections are high-efficiency mono-like cells (but square instead of "clipped"), while the outer edges are sold as conventional poly. This production method results in mono-like cells at poly-like prices.

Thin Film

Thin-film technologies reduce the amount of active material in a cell. Most designs sandwich active material between two panes of glass. Since silicon solar panels only use one pane of glass, thin film panels are approximately twice as heavy as crystalline silicon panels, although they have a smaller ecological impact (determined from life cycle analysis).

Cadmium Telluride

Cadmium telluride is the only thin film material so far to rival crystalline silicon in cost/watt. However cadmium is highly toxic and tellurium (anion: "telluride") supplies are limited. The cadmium present in the cells would be toxic if released. However, release is impossible during normal operation of the cells and is unlikely during fires in residential roofs. A square meter of CdTe contains approximately the same amount of Cd as a single C cell nickel-cadmium battery, in a more stable and less soluble form.

Copper Indium Gallium Selenide

Copper indium gallium selenide (CIGS) is a direct band gap material. It has the highest efficiency (~20%) among all commercially significant thin film materials. Traditional methods of fabrication involve vacu-um processes including co-evaporation and sputtering. Recent developments at IBM and Nanosolar attempt to lower the cost by using non-vacuum solution processes.

Silicon Thin Film

Silicon thin-film cells are mainly deposited by chemical vapor deposition (typically plasma-enhanced, PE-CVD) from silane gas and hydrogen gas. Depending on the deposition parameters, this can yield amorphous silicon (a-Si or a-Si:H), protocrystalline silicon or nanocrystalline silicon (nc-Si or nc-Si:H), also called microcrystalline silicon.

Amorphous silicon is the most well-developed thin film technology to-date. An amorphous silicon (a-Si) solar cell is made of non-crystalline or microcrystalline silicon. Amorphous silicon has a higher bandgap (1.7 eV) than crystalline silicon (c-Si) (1.1 eV), which means it absorbs the visible part of the solar spectrum more strongly than the higher power density infrared portion of the spectrum. The production of a-Si thin film solar cells uses glass as a substrate and deposits a very thin layer of silicon by plasma-enhanced chemical vapor deposition (PECVD).

Protocrystalline silicon with a low volume fraction of nanocrystalline silicon is optimal for high open circuit voltage. Nc-Si has about the same bandgap as c-Si and nc-Si and a-Si can advantageously be combined in thin layers, creating a layered cell called a tandem cell. The top cell in a-Si absorbs the visible light and leaves the infrared part of the spectrum for the bottom cell in nc-Si.

Gallium Arsenide Thin Film

The semiconductor material Gallium arsenide (GaAs) is also used for single-crystalline thin film solar cells. Although GaAs cells are very expensive, they hold the world's record in efficiency for a single-junction solar cell at 28.8%. GaAs is more commonly used in multijunction photovoltaic cells for concentrated photovoltaics (CPV, HCPV) and for solar panels on spacecrafts, as the industry favours efficiency over cost for space-based solar power.

Multijunction Cells

Dawn's 10 kW triple-junction gallium arsenide solar array at full extension

Multi-junction cells consist of multiple thin films, each essentially a solar cell grown on top of another, typically using metalorganic vapour phase epitaxy. Each layers has a different band gap energy to allow it to absorb electromagnetic radiation over a different portion of the spectrum. Multi-junction cells were originally developed for special applications such as satellites and space exploration, but are now used increasingly in terrestrial concentrator photovoltaics (CPV), an emerging technology that uses lenses and curved mirrors to concentrate sunlight onto small but highly efficient multi-junction solar cells. By concentrating sunlight up to a thousand times, *High concentrated photovoltaics (HCPV)* has the potential to outcompete conventional solar PV in the future.

Tandem solar cells based on monolithic, series connected, gallium indium phosphide (GaInP), gallium arsenide (GaAs), and germanium (Ge) p–n junctions, are increasing sales, despite cost pressures. Between December 2006 and December 2007, the cost of 4N gallium metal rose from about $350 per kg to $680 per kg. Additionally, germanium metal prices have risen substantially to $1000–1200 per kg this year. Those materials include gallium (4N, 6N and 7N Ga), arsenic (4N, 6N and 7N) and germanium, pyrolitic boron nitride (pBN) crucibles for growing crystals, and boron oxide, these products are critical to the entire substrate manufacturing industry.

A triple-junction cell, for example, may consist of the semiconductors: GaAs, Ge, and GaInP2. Triple-junction GaAs solar cells were used as the power source of the Dutch four-time World Solar Challenge winners Nuna in 2003, 2005 and 2007 and by the Dutch solar cars Solutra (2005), Twente One (2007) and 21Revolution (2009). GaAs based multi-junction devices are the most efficient solar cells to date. On 15 October 2012, triple junction metamorphic cells reached a record high of 44%.

Research in Solar Cells

Perovskite Solar Cells

Perovskite solar cells are solar cells that include a perovskite-structured material as the active layer. Most commonly, this is a solution-processed hybrid organic-inorganic tin or lead halide based material. Efficiencies have increased from below 5% at their first usage in 2009 to over 20% in 2014, making them a very rapidly advancing technology and a hot topic in the solar cell field. Perovskite solar cells are also forecast to be extremely cheap to scale up, making them a very attractive option for commercialisation.

Liquid Inks

In 2014, researchers at California NanoSystems Institute discovered using kesterite and perovskite improved electric power conversion efficiency for solar cells.

Upconversion and Downconversion

Photon upconversion is the process of using two low-energy (*e.g.*, infrared) photons to produce one higher energy photon; downconversion is the process of using one high energy photon (*e.g.*,, ultraviolet) to produce two lower energy photons. Either of these techniques could be used to produce higher efficiency solar cells by allowing solar photons to be more efficiently used. The difficulty, however, is that the conversion efficiency of existing phosphors exhibiting up- or down-conversion is low, and is typically narrow band.

One upconversion technique is to incorporate lanthanide-doped materials (Er^{3+}, Yb^{3+}, Ho^{3+} or a combination), taking advantage of their luminescence to convert infrared radiation to visible light. Upconversion process occurs when two infrared photons are absorbed by rare-earth ions to generate a (high-energy) absorbable photon. As example, the energy transfer upconversion process (ETU), consists in successive transfer processes between excited ions in the near infrared. The upconverter material could be placed below the solar cell to absorb the infrared light that passes through the silicon. Useful ions are most commonly found in the trivalent state. Er^{+} ions have been the most used. Er^{3+} ions absorb solar radiation around 1.54 μm. Two Er^{3+} ions that have absorbed this radiation can interact with each other through an upconversion process. The excited ion emits light above the Si bandgap that is absorbed by the solar cell and creates an additional electron–hole pair that can generate current. However, the increased efficiency was small. In addition, fluoroindate glasses have low phonon energy and have been proposed as suitable matrix doped with Ho^{3+} ions.

Light-absorbing Dyes

Dye-sensitized solar cells (DSSCs) are made of low-cost materials and do not need elaborate manufacturing equipment, so they can be made in a DIY fashion. In bulk it should be significantly less expensive than older solid-state cell designs. DSSC's can be engineered into flexible sheets and although its conversion efficiency is less than the best thin film cells, its price/performance ratio may be high enough to allow them to compete with fossil fuel electrical generation.

Typically a ruthenium metalorganic dye (Ru-centered) is used as a monolayer of light-absorbing material. The dye-sensitized solar cell depends on a mesoporous layer of nanoparticulate titanium dioxide to greatly amplify the surface area (200–300 m^2/g TiO2, as compared to approximately 10 m^2/g of flat single crystal). The photogenerated electrons from the light absorbing dye are passed on to the n-type TiO2 and the holes are absorbed by an electrolyte on the other side of the dye. The circuit is completed by a redox couple in the electrolyte, which can be liquid or solid. This type of cell allows more flexible use of materials and is typically manufactured by screen printing or ultrasonic nozzles, with the potential for lower processing costs than those used for bulk solar cells. However, the dyes in these cells also suffer from degradation under heat and UV light and the cell casing is difficult to seal due to the solvents used in assembly. The first commercial shipment of DSSC solar modules occurred in July 2009 from G24i Innovations.

Quantum Dots

Quantum dot solar cells (QDSCs) are based on the Gratzel cell, or dye-sensitized solar cell architecture, but employ low band gap semiconductor nanoparticles, fabricated with crystallite sizes small enough to form quantum dots (such as CdS, CdSe, Sb 2S 3, PbS, etc.), instead of organic or organometallic dyes as light absorbers. QD's size quantization allows for the band gap to be tuned by simply changing particle size. They also have high extinction coefficients and have shown the possibility of multiple exciton generation.

In a QDSC, a mesoporous layer of titanium dioxide nanoparticles forms the backbone of the cell, much like in a DSSC. This TiO 2 layer can then be made photoactive by coating with semiconductor quantum dots using chemical bath deposition, electrophoretic deposition or successive ionic layer adsorption and reaction. The electrical circuit is then completed through the use of a liquid or solid redox couple. The efficiency of QDSCs has increased to over 5% shown for both liquid-junction and solid state cells. In an effort to decrease production costs, the Prashant Kamat research group demonstrated a solar paint made with TiO 2 and CdSe that can be applied using a one-step method to any conductive surface with efficiencies over 1%.

Organic/Polymer Solar Cells

Organic solar cells and polymer solar cells are built from thin films (typically 100 nm) of organic semiconductors including polymers, such as polyphenylene vinylene and small-molecule compounds like copper phthalocyanine (a blue or green organic pigment) and carbon fullerenes and fullerene derivatives such as PCBM.

They can be processed from liquid solution, offering the possibility of a simple roll-to-roll printing process, potentially leading to inexpensive, large-scale production. In addition, these cells could be beneficial for some applications where mechanical flexibility and disposability are important. Current cell efficiencies are, however, very low, and practical devices are essentially non-existent.

Energy conversion efficiencies achieved to date using conductive polymers are very low compared to inorganic materials. However, Konarka Power Plastic reached efficiency of 8.3% and organic tandem cells in 2012 reached 11.1%.

The active region of an organic device consists of two materials, one electron donor and one electron acceptor. When a photon is converted into an electron hole pair, typically in the donor material, the charges tend to remain bound in the form of an exciton, separating when the exciton diffuses to the donor-acceptor interface, unlike most other solar cell types. The short exciton diffusion lengths of most polymer systems tend to limit the efficiency of such devices. Nanostructured interfaces, sometimes in the form of bulk heterojunctions, can improve performance.

In 2011, MIT and Michigan State researchers developed solar cells with a power efficiency close to 2% with a transparency to the human eye greater than 65%, achieved by selectively absorbing the ultraviolet and near-infrared parts of the spectrum with small-molecule compounds. Researchers at UCLA more recently developed an analogous polymer solar cell, following the same approach, that is 70% transparent and has a 4% power conversion efficiency. These lightweight, flexible cells can be produced in bulk at a low cost and could be used to create power generating windows.

In 2013, researchers announced polymer cells with some 3% efficiency. They used block copolymers, self-assembling organic materials that arrange themselves into distinct layers. The research focused on P3HT-b-PFTBT that separates into bands some 16 nanometers wide.

Adaptive Cells

Adaptive cells change their absorption/reflection characteristics depending to respond to environmental conditions. An adaptive material responds to the intensity and angle of incident light. At the part of the cell where the light is most intense, the cell surface changes from reflective to adaptive, allowing the light to penetrate the cell. The other parts of the cell remain reflective increasing the retention of the absorbed light within the cell.

In 2014 a system that combined an adaptive surface with a glass substrate that redirect the absorbed to a light absorber on the edges of the sheet. The system also included an array of fixed lenses/mirrors to concentrate light onto the adaptive surface. As the day continues, the concentrated light moves along the surface of the cell. That surface switches from reflective to adaptive when the light is most concentrated and back to reflective after the light moves along.

Surface Texturing

For the past years, researchers have been trying to reduce the price of solar cells while maximizing efficiency. Thin-film solar cell is a cost-effective second generation solar cell with much reduced thickness at the expense of light absorption efficiency. Efforts to maximize light absorption efficiency with reduced thickness have been made. Surface texturing is one of techniques used to reduce optical losses to maximize light absorbed. Currently, surface texturing techniques on silicon photovoltaics are drawing much attention. Surface texturing could be done in multiple ways. Etching single crystalline silicon substrate can produce randomly distributed square based pyramids on the surface using anisotropic etchants. It could also be etched downward to produce nanoscale inverted pyramids. Multicrystalline silicon wafers, though less effective, can be textured through a photolithographic texturing method. Incident light rays onto a textured surface do not reflect back out to the air as opposed to rays onto a flat surface. Rather some light rays are bounced back onto the other surface again due to the geometry of the surface. This process significantly improves light to electricity conversion efficiency, due to increased light absorption. In 2012, researchers at MIT reported that c-Si films textured with nanoscale inverted pyramids could achieve light absorption comparable to 30 times thicker planar c-Si. In combination with anti-reflective coating, surface texturing technique can effectively trap light rays within a thin film silicon solar cell. Consequently, required thickness for solar cells decreases with the increased absorption of light rays.

Manufacture

Solar cells share some of the same processing and manufacturing techniques as other semiconductor devices. However, the stringent requirements for cleanliness and quality control of semiconductor fabrication are more relaxed for solar cells, lowering costs.

Early solar-powered calculator

Polycrystalline silicon wafers are made by wire-sawing block-cast silicon ingots into 180 to 350 micrometer wafers. The wafers are usually lightly p-type-doped. A surface diffusion of n-type dopants is performed on the front side of the wafer. This forms a p–n junction a few hundred nanometers below the surface.

Anti-reflection coatings are then typically applied to increase the amount of light coupled into the solar cell. Silicon nitride has gradually replaced titanium dioxide as the preferred material, because of its excellent surface passivation qualities. It prevents carrier recombination at the cell surface. A layer several hundred nanometers thick is applied using PECVD. Some solar cells have textured front surfaces that, like anti-reflection coatings, increase the amount of light reaching the wafer. Such surfaces were first applied to single-crystal silicon, followed by multicrystalline silicon somewhat later.

A full area metal contact is made on the back surface, and a grid-like metal contact made up of fine "fingers" and larger "bus bars" are screen-printed onto the front surface using a silver paste. This is an evolution of the so-called "wet" process for applying electrodes, first described in a US patent filed in 1981 by Bayer AG. The rear contact is formed by screen-printing a metal paste, typically aluminium. Usually this contact covers the entire rear, though some designs employ a grid pattern. The paste is then fired at several hundred degrees Celsius to form metal electrodes in ohmic contact with the silicon. Some companies use an additional electro-plating step to increase efficiency. After the metal contacts are made, the solar cells are interconnected by flat wires or metal ribbons, and assembled into modules or "solar panels". Solar panels have a sheet of tempered glass on the front, and a polymer encapsulation on the back.

Manufacturers and Certification

National Renewable Energy Laboratory tests and validates solar technologies. Three reliable groups certify solar equipment: UL and IEEE (both U.S. standards) and IEC.

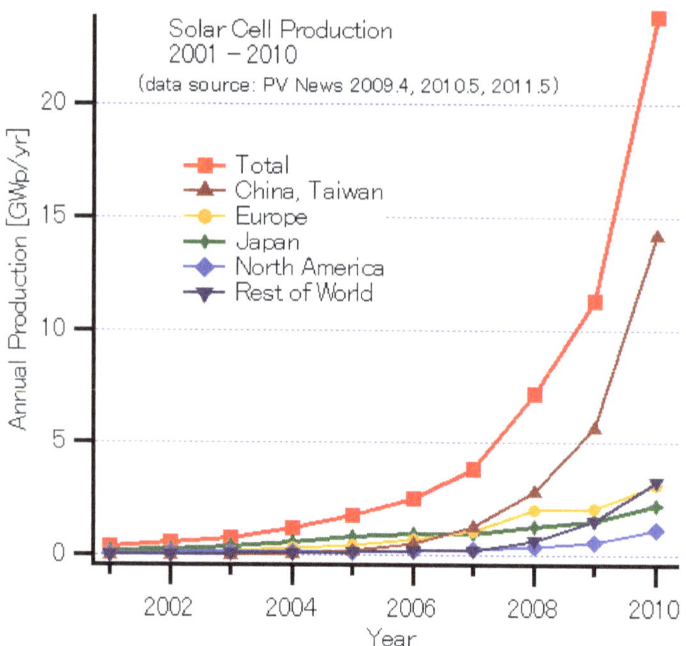

Solar cell production by region

Solar cells are manufactured in volume in Japan, Germany, China, Taiwan, Malaysia and the United States, whereas Europe, China, the U.S., and Japan have dominated (94% or more as of 2013) in installed systems. Other nations are acquiring significant solar cell production capacity.

Global PV cell/module production increased by 10% in 2012 despite a 9% decline in solar energy investments according to the annual "PV Status Report" released by the European Commission's Joint Research Centre. Between 2009 and 2013 cell production has quadrupled.

China

Due to heavy government investment, China has become the dominant force in solar cell manufacturing. Chinese companies produced solar cells/modules with a capacity of ~23 GW in 2013 (60% of global production).

Malaysia

In 2014, Malaysia was the world's third largest manufacturer of photovoltaics equipment, behind China and the European Union.

United States

Solar cell production in the U.S. has suffered due to the global financial crisis, but recovered partly due to the falling price of quality silicon.

Photographic Plate

AGFA photographic plates, 1880

Mimosa Panchroma-Studio-Antihalo Panchromatic glass plates, 9 x 12cm, Mimosa A.-G. Dresden

Negative plate

Photographic plates preceded photographic film as a capture medium in photography. The light-sensitive emulsion of silver salts was coated on a glass plate, typically thinner than common window glass, instead of a clear plastic film. Glass plates were far superior to film for research-quality imaging because they were extremely stable and less likely to bend or distort, especially in large-format frames for wide-field imaging. Early plates used the wet collodion process. The wet plate process was replaced late in the 19th century by gelatin dry plates. Glass plate photographic material largely faded from the consumer market in the early years of the 20th century, as more convenient and less fragile films were increasingly adopted. However, photographic plates were reportedly still being used by one photography business in London until the 1970s, and they were in wide use by the professional astronomical community as late as the 1990s. Workshops on the use of glass plate photography as an alternative medium or for artistic use are still being conducted.

A view camera nicknamed "The Mammoth" weighing 1,400 pounds (640 kg) was built by George R. Lawrence in 1899, specifically to photograph "The Alton Limited" train owed by the Chicago & Alton Railway, took photographs on a single glass plate measuring 8 feet (2.4 m) x 4.5 feet (1.4 m).

Scientific Uses

Astronomy

Many famous astronomical surveys were taken using photographic plates, including the first Palomar Observatory Sky Survey (POSS) of the 1950s, the follow-up POSS-II survey of the 1990s, and the UK Schmidt survey of southern declinations. A number of observatories, including Harvard College and Sonneberg, maintain large archives of photographic plates, which are used primarily for historical research on variable stars.

Many solar system objects were discovered by using photographic plates, superseding earlier visual methods. Discovery of minor planets using photographic plates was pioneered by Max Wolf beginning with his discovery of 323 Brucia in 1891. The first natural satellite discovered using photographic plates was Phoebe in 1898. Pluto was discovered using photographic plates in a blink comparator; its moon Charon was discovered 48 years later in 1978 by U.S. Naval Observatory astronomer James W. Christy by carefully examining a bulge in Pluto's image on a photographic plate.

Glass-backed plates, rather than film, were generally used in astronomy because they do not shrink or deform noticeably in the development process or under environmental changes. Several important applications of astrophotography, including astronomical spectroscopy and astrometry, continued using plates until digital imaging improved to the point where it could outmatch photographic results. Kodak and other manufacturers discontinued producing most kinds of plates as the market for them dwindled between 1980 and 2000, terminating most remaining astronomical use, including for sky surveys.

Physics

Photographic plates were also an important tool in early high-energy physics, as they get blackened by ionizing radiation. For example, in the 1910s, Victor Franz Hess discovered cosmic radiation as it left traces on stacks of photographic plates, which he left for that purpose on high mountains or sent into the even higher atmosphere using balloons.

Electron Microscopy

Photographic emulsions were originally coated on thin glass plates for imaging with electron microscopes, which provided a more rigid, stable and flatter plane compared to plastic films. Beginning in the 1970s, high-contrast, fine grain emulsions coated on thicker plastic films manufactured by Kodak, Ilford and DuPont replaced glass plates. These films have largely been replaced by digitally imaging technologies.

Medical Imaging

The sensitivity of certain types of photographic plates to ionizing radiation (usually X-rays) is also useful in medical imaging and material science applications, although they have been largely replaced with reusable and computer readable image plate detectors and other types of X-ray detectors.

Decline

The earliest flexible films of the late 1880s were sold for amateur use in medium-format cameras. The plastic was not of very high optical quality and tended to curl and otherwise not provide as desirably flat a support surface as a sheet of glass. Initially, a transparent plastic base was more expensive to produce than glass. Quality was eventually improved, manufacturing costs came down, and most amateurs gladly abandoned plates for films. After large-format high quality cut films for professional photographers were introduced in the late 1910s, the use of plates for ordinary photography of any kind became increasingly rare.

The persistent use of plates in astronomical and other scientific applications started to decline in the early 1980s as they were gradually replaced by charge-coupled devices (CCDs), which also provide outstanding dimensional stability. CCD cameras have several advantages over glass plates, including high efficiency, linear light response, and simplified image acquisition and processing. However, even the largest CCD formats (e.g., 8192x8192 pixels) still do not have the detecting area and resolution of most photographic plates, which has forced modern survey cameras to use large CCD arrays to obtain the same coverage.

The manufacture of photographic plates has been discontinued by Kodak, Agfa and other widely known traditional makers. Eastern European sources have subsequently catered to the minimal remaining demand, practically all of it for use in holography, which requires a recording medium with a large surface area and a submicroscopic level of resolution that currently (2014) available electronic image sensors cannot provide. In the realm of traditional photography, a small number of historical process enthusiasts make their own wet or dry plates from raw materials and use them in vintage large-format cameras.

Preservation

Several institutions have established archives to preserve photographic plates and prevent their valuable historical information from being lost. The emulsion on the plate can deteriorate, in addition the glass plate medium is fragile and prone to cracking if not stored correctly.

Historical Archives

The United States Library of Congress has a large collection of both wet and dry plate photographic negatives, dating from the 1855 through 1900, over 7500 of which have been digitized from the period 1861 to 1865. The George Eastman Museum holds an extensive collection of photographic plates.In 1955, wet plate negatives measuring 4 feet 6 inches (1.37 m) x 3 feet 2 inches (0.97 m) were reported to have been discovered in 1951 as part of the Holtermann Collection. These purportedly were the largest glass negatives discovered at that time. These images were taken in 1875 by Charles Bayliss and formed the "Shore Tower" panorama of Sydney Harbour. Albumen contact prints made from these negatives are in the holdings of the Holtermann Collection, the negatives are listed among the current holdings of the Collection.

Scientific Archives

Preservation of photographic plates is a particular need in astronomy, where changes often occur slowly and the plates represent irreplaceable records of the sky and astronomical objects that extend back over 100 years. An example of an astronomical plate archive is the Astronomical Photographic Data Archive (APDA) at the Pisgah Astronomical Research Institute (PARI). APDA was created in response to recommendations of a group of international scientists who gathered in 2007 to discuss how to best preserve astronomical plates. The discussions revealed that some observatories no longer could maintain their plate collections and needed a place to archive them. APDA is dedicated to housing and cataloging unwanted plates, with the goal to eventually catalog the plates and create a database of images that can be accessed via the Internet by the global community of scientists, researchers and students. APDA now has a collection of more than 200,000 photographic images from over 40 observatories that are housed in a secure building with environmental control. The facility possesses several plate scanners, including two high-precision ones, GAMMA I and GAMMA II, built for NASA and the Space Telescope Science Institute (STScI) and used by a team under the leadership of the late Dr. Barry Lasker to develop the Guide Star Catalog and Digitized Sky Survey that are used to guide and direct the Hubble Space Telescope. APDA's networked storage system can store and analyze more than 100 terabytes of data.

References

- Johnstone, B. (1999). We Were Burning: Japanese Entrepreneurs and the Forging of the Electronic Age. New York: Basic Books. ISBN 0-465-09117-2.
- Sze, S. M.; Ng, Kwok K. (2007). Physics of semiconductor devices (3 ed.). John Wiley and Sons. ISBN 978-0-471-14323-9. Chapter 13.6.

- Albert J. P. Theuwissen (1995). Solid-State Imaging With Charge-Coupled Devices. Springer. pp. 177–180. ISBN 9780792334569.
- Gevorkian, Peter (2007). Sustainable energy systems engineering: the complete green building design resource. McGraw Hill Professional. ISBN 978-0-07-147359-0.
- Tsokos, K. A. (28 January 2010). Physics for the IB Diploma Full Colour. Cambridge University Press. ISBN 978-0-521-13821-5.
- Williams, Neville (2005). Chasing the Sun: Solar Adventures Around the World. New Society Publishers. p. 84. ISBN 9781550923124.
- Kim, D.S.; et al. (18 May 2003). "String ribbon silicon solar cells with 17.8% efficiency" (PDF). Proceedings of 3rd World Conference on Photovoltaic Energy Conversion, 2003. 2: 1293–1296. ISBN 4-9901816-0-3.
- Pearce, J.; Lau, A. (2002). "Net Energy Analysis for Sustainable Energy Production from Silicon Based Solar Cells". Solar Energy (PDF). p. 181. doi:10.1115/SED2002-1051. ISBN 0-7918-1689-3.
- Dykstra, Michael J.; Reuss, Laura E. (2003). Biological electron microscopy : theory, techniques, and troubleshooting (2nd ed.). New York, NY: Kluwer Academic. p. 194. ISBN 978-0306477492. Retrieved 21 January 2016.
- "Harrow Photos - History of the Hills & Saunders Photographic Collection". Harrow School. Archived from the original on 17 April 2009. Retrieved 8 February 2016.
- "The Largest Photograph in the World of the Handsomest Train in the World" (PDF). Chicago & Alton Railway. Retrieved 30 January 2016.
- "Charon Discovery Image - Galleries - NASA Solar System Exploration". NASA Solar System Exploration. Retrieved 21 January 2016.
- "Australia's Holtermann collection of wet plate negatives" (PDF). Journal of Photography of the George Eastman House. 4 (3): 6–8. March 1955. Retrieved 23 March 2016.
- "Panorama of Sydney and the Harbour, New South Wales". Art Gallery of New South Wales. Retrieved 24 March 2016.
- "Three glass plate negatives of Sydney Harbour from the Holtermann residence, St. Leonards". State Library of New South Wales Catalogue. Retrieved 7 April 2016.

Permissions

All chapters in this book are published with permission under the Creative Commons Attribution Share Alike License or equivalent. Every chapter published in this book has been scrutinized by our experts. Their significance has been extensively debated. The topics covered herein carry significant information for a comprehensive understanding. They may even be implemented as practical applications or may be referred to as a beginning point for further studies.

We would like to thank the editorial team for lending their expertise to make the book truly unique. They have played a crucial role in the development of this book. Without their invaluable contributions this book wouldn't have been possible. They have made vital efforts to compile up to date information on the varied aspects of this subject to make this book a valuable addition to the collection of many professionals and students.

This book was conceptualized with the vision of imparting up-to-date and integrated information in this field. To ensure the same, a matchless editorial board was set up. Every individual on the board went through rigorous rounds of assessment to prove their worth. After which they invested a large part of their time researching and compiling the most relevant data for our readers.

The editorial board has been involved in producing this book since its inception. They have spent rigorous hours researching and exploring the diverse topics which have resulted in the successful publishing of this book. They have passed on their knowledge of decades through this book. To expedite this challenging task, the publisher supported the team at every step. A small team of assistant editors was also appointed to further simplify the editing procedure and attain best results for the readers.

Apart from the editorial board, the designing team has also invested a significant amount of their time in understanding the subject and creating the most relevant covers. They scrutinized every image to scout for the most suitable representation of the subject and create an appropriate cover for the book.

The publishing team has been an ardent support to the editorial, designing and production team. Their endless efforts to recruit the best for this project, has resulted in the accomplishment of this book. They are a veteran in the field of academics and their pool of knowledge is as vast as their experience in printing. Their expertise and guidance has proved useful at every step. Their uncompromising quality standards have made this book an exceptional effort. Their encouragement from time to time has been an inspiration for everyone.

The publisher and the editorial board hope that this book will prove to be a valuable piece of knowledge for students, practitioners and scholars across the globe.

Index

www.ingramcontent.com/pod-product-compliance
Lightning Source LLC
Chambersburg PA
CBHW061931190326
41458CB00009B/2715

* 9 7 8 1 6 3 5 4 9 2 1 9 4 *